Water Policy, Tourism, and Recreation

Water Policy, Tourism, and Recreation

Lessons from Australia

EDITED BY

Lin Crase and Sue O'Keefe

Routledge
Taylor & Francis Group

LONDON AND NEW YORK

First published 2011 by RFF Press

2 Park Square, Milton Park, Abingdon, Oxfordshire OX14 4RN
52 Vanderbilt Avenue, New York, NY 10017

Routledge is an imprint of the Taylor & Francis Group, an informa business

First issued in paperback 2018

British Library Cataloguing in Publication Data
A catalogue record for this book is available from the British Library

Library of Congress Cataloging-in-Publication Data
Water policy, tourism, and recreation : lessons from Australia / edited By Lin Crase and Suzanne
O'Keefe.
    p. cm.
 Includes bibliographical references and index.
 ISBN 978-1-61726-087-2 (hardback)
1. Tourism–Environmental aspects–Australia. 2. Water–Government policy–Australia. I. Crase,
Lin. II. O'Keefe, Suzanne.
 G156.5.E58W37 2011
 333.91–dc22

2010049787

ISBN: 978-1-61726-087-2 (hbk)
ISBN: 978-1-138-38005-9 (pbk)

Copyedited by Joyce Bond
Typeset by OKS Prepress Services
Cover design by Maggie Powell

About Resources for the Future *and* RFF Press

Resources for the Future (RFF) improves environmental and natural resource policymaking worldwide through independent social science research of the highest caliber. Founded in 1952, RFF pioneered the application of economics as a tool for developing more effective policy about the use and conservation of natural resources. Its scholars continue to employ social science methods to analyze critical issues concerning pollution control, energy policy, land and water use, hazardous waste, climate change, biodiversity, and the environmental challenges of developing countries.

RFF Press supports the mission of RFF by publishing book-length works that present a broad range of approaches to the study of natural resources and the environment. Its authors and editors include RFF staff, researchers from the larger academic and policy communities, and journalists. Audiences for publications by RFF Press include all of the participants in the policymaking process—scholars, the media, advocacy groups, NGOs, professionals in business and government, and the public. RFF Press is an imprint of **Earthscan,** a global publisher of books and journals about the environment and sustainable development.

The RFF Press Water Policy Series

Books in the *RFF Press Water Policy Series* are intended to be accessible to a broad range of scholars, practitioners, policymakers, and general readers. Each book focuses on critical issues in water policy with the mission to draw upon and integrate the best scholarly and professional expertise concerning the physical, ecological, economic, institutional, political, legal, and social dimensions of water use. The interdisciplinary approach of the series, along with an emphasis on real world situations and on problems and challenges that recur globally, are intended to enhance our ability to apply the full body of knowledge that we have about water resources—at local, country, regional, and international levels.

We welcome new contributions to the series. For editorial queries about the *RFF Press Water Policy Series*, please write to *waterpolicy@rff.org*.

Contents

Figures, Tables, and Boxes

FIGURES

TABLES

BOXES

Contributors

Sabine Albouy
Student Intern CSIRO Ecosystem Science
Agro Paris Tech

Dr. May Carter
School of Natural Sciences
Edith Cowan University

Dr. Bethany Cooper
Associate Lecturer
Regional School of Business
Faculty of Law and Management
La Trobe University

Lin Crase
Professor of Applied Economics
Executive Director
Albury-Wodonga Campus
La Trobe University

Brian Dollery
Professor of Economics
Director of the UNE Centre for Local Government
School of Business, Economics and Public Policy
University of New England

Dr. Ronlyn Duncan
Lecturer in Water Management
Lincoln University
Christchurch, New Zealand

Dr. Ben Gawne
Director
The Murray-Darling Freshwater Research Centre
La Trobe University

Fiona Haslam McKenzie
Professorial Fellow
Curtin Graduate School of Business
Curtin University

Dr. Darla Hatton MacDonald
Senior Research Scientist, CSIRO Ecosystem Science
Adjunct Research Fellow, Charles Sturt University

Simon Hone
Senior Research Economist, Productivity Commission
PhD Student, La Trobe University

Dr. Pierre Horwitz
Associate Professor and Postgraduate Coordinator
School of Natural Sciences
Edith Cowan University

Dr. Michael Hughes
Senior Research Fellow
Curtin Sustainable Tourism Centre
Curtin University

Colin Ingram
Director
Resolve Global Pty Ltd

Glen Jones
General Manager of the Boating Industry Association of South Australia Inc.

Dr. Sue O'Keefe
Associate Professor and Associate Head
Regional School of Business
Faculty of Law and Management
La Trobe University

Audrey Rimbaud
Student Intern CSIRO Ecosystem Science
Agro Paris Tech

Professor David G Simmons
Director Research Strategy and Development
Research and Commercialisation Office
Professor of Tourism
Faculty of Environment, Society and Design
Lincoln University, New Zealand

Dr. Sorada Tapsuwan
Research Scientist, CSIRO Ecosystem Science

Foreword

"The state of the art [of water management] is always provisional ... something that historians know well but hydrological engineers found harder to accept." So Harvard historian David Blackbourn, in *The Conquest of Nature: Water, Landscape, and the Making of Modern Germany* (2006), describes the evolution of water management over 300 years in Prussia, from a starting point where Dutch engineers helped drain a swamp and made the land livable to an endpoint where Germany's Greens defend the maintenance of "the natural landscape" today.

Stimulated by the National Competition Policy, over the past two decades Australia has undertaken the most ambitious and successful set of water reforms in the world. Against the backdrop of a drought of unprecedented brutality, the reforms have proved to be spectacularly technically successful (with minor drops in the value of agricultural output), yet hugely politically vulnerable. This paradox seems to have two causes. First, in the modern world, if anything is not right—and the state of the environment in southeast and southwest Australia certainly has not been right—then popular sentiment is that humans must have done something wrong. And second, although a perusal of the Australian section of any bookstore in the country will show that images of rugged farmers remain a powerful national symbol, agriculture now plays a much smaller demographic, economic, and value role in Australian society. A corollary is that there is a rising concern with the environment and with services to which the environment contributes, including tourism and recreation.

Australia is now in the midst of a massive rethink of its water management model. As is natural, but dangerous, there is a tendency to rely on (again in Blackbourn's words) "a series of confident prescriptions ... [that] promises to turn the trick and finally overcome the ignorance, or engineering mistakes, or political constraints of earlier generations." The legislative manifestation of this rethink is the Water Act of 2007, which has several notable features. First, it uses an environmental fig leaf—the Ramsar Convention on Wetlands—for the federal government to take over water management functions that are constitutionally in state hands. Second, it gives absolute priority to environmental uses, allowing humans to do their best with what is left. And third, it gives pride of place to "best available science" as the oracle that will tell what the environment needs.

Whether these reforms will build on the remarkable foundation laid by the generation of reforms stimulated by the National Competition Policy is a big question, as is the judgment that posterity will pass on this generation of reforms. But what is ineluctable is that Australian society has "moved on," giving much higher priority than did previous, poorer generations to the environment and related issues.

Among these related issues are two—recreation and tourism—that are of great and growing value to a society in which the right to leisure is of large and increasing importance. This book sets its sights squarely on these two issues. Building heavily on the achievements—and challenges—of the National Competition Policy generation of water reforms, an eminent group of authors expands the existing framework so that tourism and recreation benefits can find their rightful place alongside the clearly economic (water for agriculture, hydropower, and industry) and clearly social (water for people) uses of water.

Against the many challenges besetting water policy formulation in Australia, a book dealing with the tourism and recreational interests in water is particularly timely. It also provides a useful context for investigating broader public policy issues that are likely to resonate with a wider audience. Accordingly, the book examines many of the political, social, economic, and other influences that have impacts on the difficult decisions surrounding the reallocation of scarce resources. This book provides a theoretical analysis of institutional lessons and delves into the nexus between the politics and knowledge of water. Using these discussions as a backdrop, it then draws lessons from a number of practical examples. It also addresses challenges of valuation in the face of changing preferences and trade-offs between ecosystem services and the management of water resources for tourism and recreation in urban settings.

The subject matter of the book is of great relevance to the evolution of water policy and practice in Australia today. And because Australian water policy has also become a global standard, the book is of high relevance to other developed countries, which face similar issues today, as well as developing countries, which will face them in the future.

John Briscoe
Gordon McKay Professor of the Practice of Environmental Engineering
Harvard University
Former Senior Water Advisor, World Bank

Acknowledgments

The research described in this book was funded by the Sustainable Tourism Cooperative Research Centre, established and supported under the Australian government's Cooperative Research Centres program. Thanks go to the members of the industry reference group who generously gave their time to the project. The reference group comprised representatives from CSIRO, Parks Victoria, Australian Anglers Association–Victorian Division, Boat Owners Association of NSW Inc., SA Tourism, Tourism Industry Council, Tasmania, Victorian Water Industry Association, Resolve Global/Murdoch University, Griffith University, and South Australian Tourism Industry Council.

Thanks also go to Seamus Bromley and Steven Gibbs for their assistance with the preparation of this manuscript.

Acronyms and Abbreviations

ABS	Australian Bureau of Statistics
ACT	Australian Capital Territory
AMIF	Australian Marine Industries Federation
AWC	Australian Wildlife Conservancy
BIASA	Boating Industry Association of South Australia
CALM	Conservation and Land Management
CAWS	Country Areas Water Supply
CBD	central business district
CE	choice experiments
CGE	computable general equilibrium
CLLMM	Coorong, Lower Lakes, and Murray Mouth
CM	choice modeling
CoPS	Centre of Policy Studies
CSIRO	Commonwealth Scientific and Industrial Research Organisation
CVM	contingent valuation method
CWA	Clean Water Act
DEC	Department of Environment and Conservation
DMC	Drought Monitoring Center
DoW	Department of Water
ENSO	El Niño/Southern Oscillation
EPA	US Environmental Protection Agency

GDP	gross domestic product
ha	hectare(s)
HPM	hedonic pricing method
IBT	inclining block tariff
IO	input–output
IPART	Independent Pricing and Review Tribunal
km	kilometer(s)
km^2	square kilometer(s)
KNP	Kruger National Park
LRMC	long-run marginal cost
LWD	large woody debris
m	meter(s)
m^2	square meter(s)
MA	Millennium Ecosystem Assessment
MDB	Murray-Darling Basin
MDBA	Murray-Darling Basin Authority
MWSSD	Metropolitan Water Supply, Sewerage and Drainage
MWWG	Murray Wetlands Working Group
NCC	National Competition Council
NIMBY	not in my backyard
NMMA	National Marine Manufacturers Association
NOAA	US National Oceanic and Atmospheric Administration
NSW	New South Wales
NWC	National Water Commission
NWI	National Water Initiative
OT	Oregon Trout
OWT	Oregon Water Trust
PDWSA	public drinking water source area
PEAC	Pacific ENSO Applications Center
QLD	Queensland
RP	revealed preference
SA	South Australia

SADC	Southern African Development Community
SARCOF	Southern African Regional Climate Outlook Forum
SD	Statistical Division
SP	stated preference
SRMC	short-run marginal cost
STS	science and technology studies
TCM	travel cost method
TERM	The Enormous Regional Model
TEV	total economic value
USAPI	United States Affiliated Pacific Islands
WA	Western Australia
WTA	willingness to accept
WTP	willingness to pay

PART I

CONTEXT, VALUES, AND TRADE-OFFS

CHAPTER I

The Policy Landscape and Challenges for Tourism and Recreation in Australia

Lin Crase, Sue O'Keefe, and David Simmons

*H*igh-quality freshwater resources are essential to both tourism and recreation, and these activities continue to gain economic prominence. It is puzzling, then, that relatively little is known about the value of water for tourism or of the extent of conflicts and complementarities with other uses of the resource. Internationally, despite tourism's significance to regions and nations, it is often absent from national policy considerations, especially in the context of water (Richter, 1983; Hall and Jenkins, 1995). Several reasons are put forward for this, including the sector's disparate nature and the perception that it comprises hedonistic and wasteful pursuits compared with the "real" work of agriculture or construction.

In Australia, the relationship between water and tourism outputs has recently been brought into sharp relief as water policy has become hotly contested. The controversy surrounding the recent release of the *Guide to the Basin Plan* is evidence of the perceived conflict between agricultural and environmental uses within the Murray-Darling Basin. The guide sets forth basinwide targets to reduce the extent of overallocation and thereby provide more water for the environment, but farmer lobbies have organized swiftly and vehemently protest that this will result in the death of regional communities within the basin. In contrast, despite the importance of water for tourism and recreation, these interests remain silent. This is particularly perplexing given that tourism accounts for almost double the employment of agriculture in the basin (ABARE 2010).

Many regional areas of Australia have a significant investment in the tourism industry. This sector employed 497,800 people in 2007–2008 (ABS 2010), and its share of the nation's gross domestic product (GDP) was 4%. This outstrips the contribution of agriculture, which sits at around 3% of GDP (ABS 2009). Nonetheless, research and policy attention has focused on the conflicting interests of consumptive users like agriculture versus the desire for restoring or maintaining nonconsumptive supplies for environmental benefit. The apparent disconnect between the economic significance of tourism and recreation and the standing of these interests in water affairs might be traced to several factors. Among these is the paucity of research into the nexus between water and tourism, especially in the Australian context.

In this book, we aim to address this research deficit and explore the complicated interrelationships between fresh water and tourism and recreation. The focus is on Australia, but we make comparisons with the experiences of other countries throughout. The book attempts to shed light on the potential complementarities and conflicts between a "more natural but variable" hydrological cycle and the activities of the tourism industry, raising issues around the scope for adjusted institutional arrangements. Matters relating to the trade-offs and valuation of water in its various uses are also addressed.

This introductory chapter sets the context of the contributions by others and is organized into four main parts. First, we provide as background a synoptic overview of the policy context in Australia.[1] This is followed by a survey of recent international and Australian research on water and tourism, from which we draw four main themes that form the organizing principles of this book. The final parts of the chapter are devoted to the aims, objectives, and structure of the manuscript.

POLICY CONTEXT IN AUSTRALIA

Australia is frequently cited as being a dry country, but it would be more accurate to note that rainfall is highly variable in both temporal and spatial terms. The important upshot from this is that the ecology of Australia is well adapted to a variable resource, although most agricultural activity has been premised on European production systems that hinge on making water more "reliable" or "secure." These two circumstances have combined to produce an irrigated agriculture sector that is costly to maintain and ecological systems that are severely threatened by the radical modification to hydrology required by irrigated agriculture.

Making streams and rivers "reliable" by regulating them carries with it a range of environmental costs. Some of these have arisen because the quantity of water is simply no longer available to support ecosystems, while structural and managerial aspects of water regulation have brought additional challenges to the natural ecology. Hillman (2009) summarizes these into categories covering impacts from interbasin transfers, physical barriers that prevent the passage of species, depressed summer water temperatures, inverted streamflow, modifications to short-term variability of flow, and the removal of various flow classes. These ecological costs have proved significant. For example, the Murray-Darling Basin Authority's Sustainable Rivers Audit recently reported that only 1 of 23 river valleys in that basin was in good condition, while 13 were in very poor condition (Davies et al., 2010).

Smith (1998) contends that Australian water storages need to be about six times larger than those found in Europe to deliver the same level of reliability.

Davidson (1966) poignantly noted that irrigated agriculture in Australia was always destined to be more costly than its counterparts in other parts of the world (or dryland agriculture, for that matter). Nevertheless, major public investments were made in regulating water resources, including the development of communal irrigation districts in most states. This was often undertaken under the guise of social policy and driven by faith in the notion of yeomanry, the perceived necessity for self-sufficiency, and the desire for closer settlement. By the end of the 1960s, the Australian water landscape was characterized by large engineering structures that stood as testament to the political enthusiasm for nation building, an irrigated agriculture sector that was continually encouraged to expand water use "for the good of the nation," and formidable water and agricultural bureaucracies charged with managing all manner of activities that might be considered part of the private domain.

Interest in reforming water management grew from two main quarters in the 1970s and 1980s. On the one hand, governments became increasingly uneasy with the notion that irrigated agriculture was subject to market failure and thus required continual state intervention and support. This was exacerbated by the maturity of the water economy, with most of the low-cost alternatives for water development already exploited (Randall, 1981). Accordingly, interest was now much greater in pricing water to recover cost, devolving responsibility for the management of irrigation assets, and clearer separation of the roles of resource steward, irrigation supplier, and regulator. On the other hand, during the same period, the environmental impacts of excessive water extraction and modified hydrology were becoming increasing evident. This included dire predictions about the impacts of salinity and major incidents exposing the deterioration of water quality, such as the blue-green algae bloom that covered more than 1,000 kilometers of the Barwon-Darling River system in October and November 1991. Thus by the beginning of the 1990s, governments in most jurisdictions faced increasing pressure to deal with these two issues.

Arguably, the greatest enthusiasm for reform was first evident in the states that share the Murray-Darling Basin, where the regulation of rivers and water resource extraction had been most pronounced. Coincidentally, these states are the nation's most populous, and the national capital also relies on water resources from the basin. The consequence has been a policy agenda heavily influenced by the status of the basin's water resources.

Against this initial background, water policy focus since the 1990s might be categorized as falling into three main phases: first, institutional and legislative adjustment with minimal sectoral reallocation; second, expanded reallocation effort with a strong focus on economic instruments; and third, a retreat to subsidy and engineering. These interventions have been circumscribed by marked international and national interest in the impacts of climate change and extended drought, particularly in the southern Murray-Darling Basin.

We now briefly describe each of these policy platforms to gain an appreciation of the sectoral influences that have shaped water policy to date and to reflect on the types of information that have had greatest impact. The purpose is to highlight the information gaps pertaining to the tourism sector.

Institutional and Legislative Adjustment with Minimal Sectoral Reallocation

The process of national water reform commenced in earnest with the introduction of the Water Reform Framework in 1994–1995. This effectively subsumed water policy into the National Competition Policy framework and encouraged compliance via the financial inducements offered to the states. The early agenda focused heavily on establishing better institutional arrangements and clarifying the principles of cost recovery and future investment. There was also support for the development of tradable water rights that were separate from land. For example, Hall et al. (1994) advocated strongly for the introduction of trade among agriculturalists in the Murray-Darling Basin, on the grounds that misallocation was costing around A\$50 million per year.[2]

The early reforms gave some emphasis to environmental claims, and some investments were made in understanding the ecological relationships between water and ecosystems during this phase, although many gaps in knowledge were identified (see, e.g., DIST 1996, 67). Most jurisdictions responded by crafting new legislation and developing plans around the mechanisms for sharing water between extractive users and the environment. Perhaps not surprisingly, most plans focused on agricultural users, who use about two-thirds of the resource nationally (ABS 2008), and lobbyists seeking to restore environmental assets. An inspection of the various reviews undertaken by the National Competition Council at the time (see, e.g., NCC 2004) shows that although progress had been made on the planning front, little genuine resource reallocation was actually being done.

Intense political lobbying was occurring on both fronts. Agriculturalists argued persuasively that the economic and social ramifications of reallocation would be severe, and some sought to cast doubt over the caliber of the scientific evidence on environmental degradation (Marohasy, 2003). However, the environmental lobby secured agreement across the Murray-Darling Basin that total water extractions should be capped at predetermined levels. This group also gained support for the Living Murray Initiative, which aimed to return 500 gigaliters to the River Murray by 2009.[3] An important force at this time was the multidisciplinary Wentworth Group of concerned scientists.

Several other noteworthy trends began to emerge at this time. First, there was a strong focus on measuring water reallocation in volumetric terms, in contrast to Hillman's (2009) typology of flows. Second, water markets were continually cited as the means of reallocating water to higher-value uses (see, e.g., Productivity Commission, 2006), although attention concentrated on shifting the resource among different forms of agriculture, and the potential trade-offs and complementarities among tourism, irrigation, and the environment were never seriously addressed. Third, government support for adjustments in agriculture to achieve "water savings" or "water use efficiency" became more apparent at this time.

Expanded Reallocation Effort with a Strong Focus on Economic Instruments

The pronouncements embodied in the National Water Initiative (NWI) of 2004 signified the commencement of this second phase. The NWI continues to be proclaimed as one of the most accomplished policy initiatives in this field for some time (Young, 2009). One of the main focuses of the NWI was on refining the entitlements to water. In many respects, this could be viewed as a response to the sequencing problems that had arisen from earlier reforms. Having opted to promote the separation of water and land, and insisting on the creation of water markets, the earlier reforms revealed significant deficiencies in the way water rights were specified. This also led to the sale and activation of many water rights that were previously dormant and thus exacerbated the overallocation problem in many catchments. In most jurisdictions, this has resulted in water rights being unbundled and the access component being specified as a variable share of a consumptive pool. These rights are then tradable, although the state continues to place strong caveats over trade that generally advantages the agricultural sector.

Accompanying the attention to markets and water rights has been a greater acknowledgment of the poor ecological status of many of the nation's river systems during this phase. The ecological research effort continued to expand, although progress and approaches varied across jurisdictions, a point taken up later in this book.

This phase was also characterized by increased interest in urban water use. For instance, projects were developed around water use efficiency in urban areas, the impacts and acceptability of recycled urban wastewater, and the requirements for making water-sensitive cities (NWC 2008). A number of studies also emerged around improving our understanding of the economic impacts of reallocating water between urban and agricultural users (see, e.g., Dwyer et al., 2005). Despite these sectoral advances, the tourism sector was conspicuous in its absence from studies of policy impacts and considerations.

The support for water reform via the public purse expanded during this phase, as did the commonwealth's influence over water affairs. The National Water Commission (NWC) was created to administer the Australian Government Water Fund. Most emphasis in funding programs was given to subsidizing irrigation upgrades, although other areas to receive support included the development of broad-based knowledge and water accounting systems, urban water systems, and water-dependent ecosystems. A review of these programs shows no support for projects specifically targeted at improving understanding of the relationship between tourism and water.

In sum, this phase of reform was characterized by more rigorous approaches to the design of economic instruments, but the impacts of these were considered almost exclusively in terms of irrigation and ecological outputs. Some interest in urban water use surfaced, but this was modestly supported by the public purse compared with other areas. The importance of water to tourism was effectively ignored. The outcomes of policy were more substantial than during the early phase as water markets became more active as a vehicle for reallocating the resource. Nevertheless, overallocation persisted at the expense of the environment in many areas.

Retreat to Subsidy and Engineering

The final phase of reform commenced with the Howard government's National Plan for Water Security and the Rudd government's Water for the Future policy.

The former related to a proposal to spend A$10 billion over 10 years, with the majority of support focused on assisting adjustments in agriculture. This was to be accompanied by the commonwealth formally assuming water management responsibilities in the Murray-Darling Basin. The current government's approach broadly mirrored this, although additional funding is now involved, and greater enthusiasm for using water markets to address overallocation is evident.

Funding for the NWC expanded during this final phase, and attention is now given to groundwater and its relationship to urban, agricultural, and ecological demands. Investments in building knowledge around agriculture and water use have continued to materialize, although they have tended to focus on analyses of the merits of water use efficiency (see, e.g., Wong, 2008) or studies concerned with the impacts of water trade on agricultural communities (e.g., NWC 2008).

On the ecological research front, jurisdictions appear to be pursuing fragmented studies, which hamper attempts to design better decision support tools to achieve environmental outcomes (Morton, 2008). This remains a formidable challenge to better policymaking (NWC 2008).

Notwithstanding these problems, at least a modicum of knowledge exists about the ecological responses to water. Similarly, there is now a body of work about the economic impacts of different water scenarios on agriculture and urban water users. Similar claims cannot be made about the tourism sector in this country. Several reasons may be advanced to explain this omission, the most obvious of which relates to the complexity and heterogeneity inherent in the sector.

BACKGROUND TO TOURISM WATER RESEARCH

When considering the relationship between tourism and fresh water, defining the tourism sector is a challenge in its own right. According to the Australian Bureau of Statistics (ABS 2003), tourism is defined according to the status of the customer rather than the nature of the goods and services produced. This means that it is not a discrete and readily discernible sector in itself, and it is also likely that the value of recreational activity is often not easily distinguished from tourist activity. Take the case of a group of anglers fishing a waterway. Typically, some of these anglers will have traveled far to undertake the activity, whereas others may live locally. Unless a methodology like travel cost method is used to value this activity, it may not be feasible or even desirable to disaggregate local demand from tourist demand.

Examination of the literature uncovers four key strands of inquiry that might be used to shape our thinking about tourism and water. These relate to the following:

- the concept of value and the conflicts and complementarities between water for tourism and recreation and for other uses;
- the importance of property rights and institutions and their impact on decisionmaking;
- current practice in the tourism sector and its relationship to water; and
- the relationship between tourist behavior and water infrastructure, such as that embodied in urban water supply.

These are the themes that form the organizing principles for this book, and they are briefly addressed below.

Production Relationships, Values, and Trade-offs

An understanding of the relationships between water inputs and the value generated by tourism and recreation is an important step in gaining policy influence. For instance, agricultural interests have been able to demonstrate the impact on irrigation outputs and then apply standard input–output models to illustrate community impacts (see, e.g., MDBMC 2002). Similarly, environmental interests are now actively involved in illustrating the "productive" impacts on ecosystems of particular water management regimes (MDBA 2009). However, this approach also relies on having some way to measure the economic value of water in its alternative uses (Ward and Michelsen, 2002, 442). From an agricultural perspective, this can be achieved by reference to water markets (at least since the mid-1990s) or product markets, but this approach has some serious limitations, particularly for other sectors. First, water markets in Australia fall well short of the competitive model (the insistence by state governments that water exports be capped from communal irrigation districts is a case in point), and second, not all water users are equipped to bid in the water market, an issue explored in greater depth in this volume. For instance, a heterogeneous sector like tourism is not always able to coordinate in a manner that would allow it to express its values through the water market. Third (and perhaps most significantly, in the context of water), water markets in Australia have been created almost exclusively around volumetric property rights, yet volumes can be largely meaningless for tourism and recreation interests, who may have more interest in flow or the timing of flows.

Some work has been done on the production relationships between tourism and water. Smith (1994) conceptualized water as a primary input used in the production of intermediate inputs, which in turn are used to generate other intermediate outputs and final outputs. Tourism constitutes multiple outputs, but the demands of tourism and recreation on water for the production of these outputs are not always complementary. For example, farm tourism centered on the consumption of irrigated wine, where water is required in the summer months, is likely to be at odds with the fishing of inland waterways, where fish species are more populous if flows are concentrated in winter and spring. This has important implications for the manner in which the collective voice of tourism and recreation might be heard in policy circles.

Although the framework of Smith (1994) provides a useful starting point, examination reveals several deficiencies. Among these, water is not always strictly a private good when used as an input in tourism and recreation. For instance, a hike in a public space might be enhanced by the presence of a pristine waterway or stream, or conversely, polluted water may have negative effects on the experience. To understand the impact of water on this "output," we must be able to value the output and to measure the marginal contribution of the water in that setting. This proves challenging when the water itself is unpriced.

Marcoullier (1998) considers natural resources, such as water, as latent primary factors of production in tourism. One of the difficulties with Marcoullier's approach is that it considers the private and public good dimensions of resource management as being at odds. Thus, as the private good components expand, the public good components contract. However, with water, complex feedback loops exist between private and public good components, especially in the context of recreation and tourism.

These complexities probably explain why relatively little theoretical or empirical work has been undertaken to expand on all components of Smith's (1994) framework. Rather, ad hoc studies have shed some light on some components of these relationships, particularly the link between the quantity of water at a particular point in space and time and the value that users place on it. International examples of this type of work are relatively common. For instance, Laitila and Paulrud (2008) recently analyzed the relationship between the removal of a dam to restore "natural" water flows and the willingness of Swedish anglers to pay for this change to occur.

In the United States, heightened concern over recreation and tourism has driven environmental regulation and legislation to protect habitat (Ribaudo et al., 1990), and the value of water for tourism and recreation has assumed a prominent role in the protracted allocation disputes that have occurred since the early 1990s. For example, a study conducted by McKean et al. (2005) in the U.S. state of Washington investigated the impact of dam breeching to protect salmon. Using the travel cost method, McKean and colleagues calculated the willingness to pay for current nonangling recreation uses and noted several complementarities among tourism, recreation, and environmental benefits.

Studies of this form are notably less common in an Australian context. However, Crase and Gillespie (2008) reported the results of an empirical analysis contrasting different water levels with recreational visits and expenditure at Lake Hume, situated at the headwater of the River Murray. Other emerging approaches include the use of hedonic pricing to tackle the relationship between recreational green space and the value of housing infrastructure. Hatton MacDonald and others in the Commonwealth Scientific and Industrial Research Organisation (CSIRO) are leading this work. Similarly, Brennan et al. (2007) used a production function for turf to estimate the financial impacts of water restrictions for the population at large, and this approach could easily be modified to focus directly on the impact on tourism assets.

Nonetheless, despite the existence of several well-established empirical methods to calculate the value of water in different contexts, these have seldom been employed to establish the value of water for tourism. Accordingly, the Australian policymaker is not well placed to meet Ward and Michelsen's imperative to develop "conceptually correct and empirically accurate" estimates of value for water to facilitate the "rational allocation of scarce water across locations, uses, users and time periods" (2002, 423).

Accordingly, a key question addressed in the first part of this book is, what is known about the value of water for tourism and recreation and, more specifically, about the conflicts and complementarities among the various uses of the water resource?

Property Rights, Institutions, and Decisionmaking

Uncertainty about the value of water in its various uses has meant that decisions about the allocation of the water resource typically fall in the political realm. Water is an integral part of the ecosystem, a natural resource and a social and economic good, and the expression of users' preferences is circumscribed by the shape of property rights and institutions. Property rights affect the behavior of actors, and therefore the crafting of property rights and water institutions is critical in decisionmaking. However, water has multiple dimensions. Quantity is the simplest and most frequently used metric, but other aspects should be considered as well. Property rights are currently expressed in quantitative terms, but high-quality water is critical for most uses, and timing is also a key element. For example, the efficacy of environmental flows depends as much on timing as it does on quantity (see, e.g., Hillman, 2009), and full reservoirs are of little use to water-skiers in the dead of winter. The building of dams represents an attempt to capture enhanced time utility of a given quantity of water (Ward and Michelsen, 2002).

To further complicate water allocation decisions, whereas other natural resources, once used, are not available for consumption by others, the use of water in one context does not necessarily preclude its use further downstream. From another perspective, allocating water to one user also has the potential to have adverse impacts on the quality, timing, reliability, and location of supplies for other uses (Frederick et al., 1996, 5).[4] Some water uses are also likely to be complementary; for example, hydroelectricity and some forms of recreation can easily coexist. The challenge is to assemble sufficient information to ensure that all material interests in water, including those of tourism and recreation, are considered in the allocation decision, or to have in place institutions that allow these interests to be expressed.

Experience in other countries has shown that tourist and recreational interests can be served through participation in markets where the institutional structure allows for it (see, e.g., King, 2004). The development of water markets in Australia raises this possibility, but given the complexities of the water resource when viewed from a tourism and recreation perspective, the question remains as to whether the current institutions and property rights can secure the best outcome. The Australian context arguably has been constrained by top-down decision-making and a proclivity for the government to meddle in the market when it comes to matters of water allocation. The second strand of inquiry addressed in this book relates to property rights and institutions: how does the crafting of property rights and related institutions constrain or influence tourism water management decisions?

Current Issues in Water Policy for Tourism and Recreation

There are significant politicoscientific lessons to be learned from analyzing specific cases that relate to water and tourism and recreation. These cases can raise a range of issues and challenges for the tourism sector and help us understand the constraints of theory. The issues related to value, trade-offs, and institutions play out in various ways as we examine specific tourism scenarios. A particularly glaring example of the conflicts among various water uses relates to the closure of water catchments located close to major centers of population. Much recreation and tourism demand arises in periurban areas, and at times, conflict arises between the uses of these areas for recreational and tourism pursuits and their status as protected areas for the preservation of water quality. Moreover, the institutions surrounding decisions to exclude access play a vital role in dealing with the attendant trade-offs. The result is that in some settings, these trade-offs potentially can be reduced by adequate institutional design.

One lesson that emerges is the importance of political persuasion in many of the decisions that surround the allocation of the water resource. This can be evidenced by examining the experiences of different groups of interested parties within the tourism and recreation sector that have been able to secure improved outcomes for their constituents. More specifically, in addition to understanding values, trade-offs, and institutions, it would appear to be valuable to consider the nexus between the knowledge and the politics of water. The conceptualization of how the tourism and recreation sector might best influence policy is critical here. More generally, this approach offers an alternative lens to conceptualize how water policy is formulated. Accordingly, the question that arises is, what does current practice tell us about policy formulation, and how can knowledge be better developed and aligned to serve the interests of competing users?

Potable Water and Tourism

The literature on the production nexus between fresh water and tourism and recreation is patchy and often site-specific, and it may ignore important feedback loops or the nuances of hydrology. The studies described above show a predilection to focus on the nonconsumptive relationship between tourism and recreation and water, but another important dimension relates to the consumptive relationships.[5] In this regard, it is worth noting that the water-using behavior of tourists has significant infrastructure and planning implications.

Internationally, these concerns have garnered a response from the research community, with most studies showing that per capita water use by tourists far exceeds that of local residents (see, e.g., Cullen et al., 2003, 18; De Stefano, 2004; Gajraj, 1981; Gössling, 2001, 2002; Stonich, 1998). Oddly enough, however, a similar level of research enthusiasm is not evident in this field in Australia. This is peculiar insomuch as the hydrological variability of this country implies even greater need for understanding on this front. These issues are addressed in Chapter 12.

Some examples of institutional and policy work can be found within the agencies and regulators concerned with water and wastewater supply (see, e.g., Westernport Water, 2009), but little of this is founded on theoretical or rigorous empirical study. The upshot is that the pragmatic approach to water pricing is not always aligned with other policy goals or the findings of research into pricing structures and tariffs (see, e.g., Crase et al., 2008). These complexities are addressed further in Chapter 13. The central question is, what are the implications of tourist water-using behavior on water infrastructure, such as that embodied in urban water supply?

OBJECTIVES OF THE BOOK

A book dealing with the tourism and recreational interests in water is timely in Australia in the context of continued competition for the scarce resource, and it also serves as a vehicle for investigating a range of important public policy issues with lessons for a wider audience.

First, research in this field provides an opportunity for analyzing the range of forces at work when dealing with the reallocation of scarce, and often overallocated, resources. These matters are brought sharply into focus in regions like the Murray-Darling Basin, where, in the context of climate change, the historical dominance of agriculture is being challenged by declining terms of trade relative to other resource claimants. The calculation of value is vital in this respect, as sensible water policy should take account of the value of water in various contexts, including nonmarket values. It should also take account of the fact that values change over time and the necessity to limit institutional impediments that would prevent water moving to reflect these changes.

Second, important institutional lessons may be gained from this analysis. Some of these relate to the property rights to water and how the shaping of property rights affects the behavior of various players. In addition, there are institutional issues that pertain to the coordination of disparate interests, as is often the case for the tourism and recreation sector, where small to medium business is the norm.

Third, significant politicoscientific lessons may also be learned from analyzing water and tourism and recreation. More specifically, we have an opportunity to consider the nexus between the knowledge and the politics of water. The conceptualization of how the tourism and recreation sector might influence policy is critical here. More generally, this approach offers an alternative lens to conceptualize how water policy is formulated.

Finally, a book in this field provides scope to analyze the behavioral dimensions of tourists and recreationists. This will provide salient lessons for a range of practitioners and policymakers.

APPROACH AND STRUCTURE

This book encapsulates the assortment of issues confronting those involved with water resource management and dealing with the changing priorities in water use and access. We seek to answer the four key questions raised above.

The book is organized into four main parts. Part I provides an overview of the context of Australian water resource management, looks at the status of tourism and recreation in the country, and highlights the complementarities and conflicts surrounding water use and access. In Chapter 2, Simon Hone describes the current state of rivers in Australia and discusses the "science of water," including matters relating to climate change, drought, ecosystems, and pollution. In Chapter 3, Darla

Hatton MacDonald, Sorada Tapsuwan, Sabine Albouy, and Audrey Rimbaud examine in detail the central concept of value, including an overview of the valuation methods currently in use and findings about the value of tourism and recreation in the Murray-Darling Basin. The notion of trade-offs is central to this section, and in Chapter 4, Pierre Horwitz and May Carter introduce a framework to conceptualize trade-offs in ecosystem services.

Institutional arrangements around water and the significance of property rights are considered in Part II. The salience of hydrology is reemphasized, with some predictive analysis of increased water variability and the lessons for institutional design developed. Lin Crase and Ben Gawne examine some of the challenges in defining property rights in Chapter 5. Australia is often regarded as being at the forefront of water management and institutional design, in spite of the limited consideration of tourism and recreation interests in water. The development of water markets and water pricing are cases in point. Against the backdrop of increasing importance of the tourism and recreation sector, both in Australia and abroad, theoretical models for increasing collaboration and cooperation are the subject of Chapter 6, by Brian Dollery and Sue O'Keefe. This theme is further developed by Sue O'Keefe and Brian Dollery in Chapter 7, which examines practical examples of collaboration such as the use of trusts.

Part III delves into the formulation of water policy, the role of science, and the political dimensions of the process. In Chapter 8, Fiona Haslam McKenzie deals with the political dimensions by close examination of a case study into the operation of the Swan River Trust in Western Australia. A further case study is considered by Michael Hughes and Colin Ingram in Chapter 9, dealing specifically with the practical trade-offs between access to urban water supplies and the requirements of recreation. In Chapter 10, Sue O'Keefe and Glen Jones highlight the endeavors of interest groups such as the boating industry to influence policy thinking. In the final chapter in this section, Chapter 11, Ronlyn Duncan draws on the three previous case study chapters to suggest a new conceptualization of the production of knowledge that is seen as having potential to increase the influence of the tourism and recreation sector in the water policy arena.

Part IV focuses on the complexities that attend the relationship between tourism and urban water demands and explores the behavioral elements of tourism activities. In Chapter 12, Bethany Cooper considers how water-using behavior differs for individuals during recreational pursuits or on vacation. This is extended to include the implications for infrastructure design, labeling, and accreditation in the sector. Lin Crase and Bethany Cooper look at the role of water pricing and water tariffs in Chapter 13. In the final chapter in the book, Chapter 14, Sue O'Keefe and Lin Crase bring together the lessons from the various contributors and review the agenda for research.

NOTES

1. For a more complete examination, see Crase (2008).
2. A$1 = US$0.9978 as of January 2011.

3. Environmentalists might question whether this was a success, as scientific evidence pointed to the need for at least 1,500 gigaliters to produce a discernible change to the environmental status of the river.

4. More recent approaches to water accounting take these complex relationships into account by conceptualizing water balance at the basin level (Molden and Sakthivadivel, 1999).

5. These two distinctions are not strictly enforceable. As noted later, the way potable water sources are managed for consumptive purposes has implications for the nonconsumptive recreational uses of water in water supply catchments. The burgeoning importance of the sector has been driven in part by rising standards of living, accompanied by a move toward service economies and an increased emphasis on environmental values within the polity.

REFERENCES

ABARE (Australian Bureau of Agricultural and Resource Economics). 2010. *Environmentally Sustainable Diversion Limits in the Murray-Darling Basin: A Socioeconomic Analysis.* Canberra: Australian Government Publishing Service.

ABS (Australian Bureau of Statistics). 2003. *Framework for Australian Tourism Statistics.* cat: 9502.0. Canberra: Australian Government Publishing Service.

———. 2008. *Year Book Australia, 2008.* cat: 1301.0. Canberra: Australian Government Publishing Service.

———. 2009. *Agriculture Statistics Collection Strategy—2008–09 and Beyond.* cat: 7105.0. Canberra: Australian Government Publishing Service.

———. 2010. *Year Book Australia, 2009–10.* cat: 1301.0. Canberra: Australian Government Publishing Service.

Brennan, D., S. Tapsuwan, and G. Ingram. 2007. The welfare costs of urban outdoor water restrictions. *Australian Journal of Agricultural and Resource Economics* 51 (3): 243–262.

Crase, L. (Eds), 2008. *Australian Water Policy: The Impact of Change and Uncertainty.* Washington, DC: RFF Publications.

Crase, L., and R. Gillespie. 2008. The impact of water quality and water level on the recreation values of Lake Hume. *Australasian Journal of Environmental Management* 15 (1): 21–29.

Crase, L., S. O'Keefe, and J. Burston. 2008. Inclining block tariffs for urban water. *Agenda* 14 (1): 69–80.

Cullen, R., J. McNicol, G. Meyer-Hubbert, D.G. Simmons, and J.R. Fairweather. 2003. *Tourism, Water and Waste in Akaroa: Implications of Tourist Demand on Infrastructure.* Report No. 38. Canterbury, New Zealand: Tourism Recreation Research and Education Centre (TRREC), Lincoln University.

Davidson, B. 1966. *The Northern Myth.* Melbourne: Melbourne University Press.

Davies, P., J. Harris, T.J. Hillman, and K.F. Walker. 2010. The sustainable rivers audit: Assessing river ecosystem health in the Murray-Darling Basin. *Marine & Freshwater Research* 61: 764–777.

De Stefano, L. 2004. *Freshwater and Tourism in the Mediterranean.* Rome: WWF Mediterranean Program.

DIST (Department of Industry, Science and Tourism). 1996. Managing Australia's inland waters: Role for science and technology. paper prepared by independent working group for the consideration of the Prime Minister. Canberra: Department of Industry Science and Tourism.

Dwyer, G., P. Loke, S. Stone, and D. Peterson. 2005. Integrating rural and urban water markets in South East Australia: Preliminary analysis. paper presented at OECD Workshop on Sustainability, Markets and Policies, November 14–18, 2005, Adelaide.

Frederick, K.D., T. VandenBerg, and J. Hanson. 1996. *Economic values of freshwater in the United States.* Washington, DC: Resources for the Future.

Gajraj, A.M. 1981. Threats to the terrestrial resources of the Caribbean. *Ambio* 10 (6): 307–311.

Gössling, S. 2001. The consequences of tourism for sustainable water use on a tropical island: Zanzibar, Tanzania. *Journal of Environmental Management* 61 (2): 179–191.

———. 2002. Global environmental consequences of tourism. *Global Environmental Change* 12 (4): 283–302.

Hall, C.M., and J. Jenkins. 1995. *Tourism and Public Policy.* London: Routledge.

Hall, N., D. Poulter, and R. Curtotti. 1994. *ABARE Model of Irrigation Farming in the Southern Murray-Darling Basin.* ABARE Research Report 94.4. Canberra: Australian Bureau of Agricultural and Resource Economics.

Hillman, T. 2009. The policy challenge of matching environmental water to ecological need. In *Policy and Strategic Behaviour in Water Resource Management,* edited by A. Dinar and J. Albiac. London: Earthscan, 109–124.

King, M. 2004. Getting our feet wet: An introduction to water trusts. *Harvard Environmental Law Review* 28: 495–534.

Laitila, T., and A. Paulrud. 2008. Anglers' valuation of water regulation dam removal for the restoration of angling conditions at Storsjo-Kapell. *Tourism Economics* 14 (2): 283–296.

Marcoullier, D. 1998. Environmental resources as latent primary factors of production in tourism: The case of forest based commercial recreation. *Tourism Economics* 4 (2): 131–145.

Marohasy, J. 2003. *Myths and the Murray: Measuring the Real State of the River Environment.* Melbourne: Institute of Public Affairs.

McKean, J.R., D. Johnson, R.G. Taylor, and R.L. Johnson. 2005. Willingness to pay for non angler recreation at the Lower Snake River Reservoirs. *Journal of Leisure Research* 37 (2): 178–194.

MDBA (Murray-Darling Basin Authority). 2009. *Issues Paper: Development of Sustainable Diversion Limits for the Murray-Darling Basin.* Canberra: Murray-Darling Basin Authority.

MDBMC (Murray-Darling Basin Ministerial Council). 2002. *The Living Murray: A Discussion Paper on Restoring the Health of the River Murray.* Canberra: Murray Darling Basin Committee.

Molden, D., and R. Sakthivadivel. 1999. Water accounting to assess use and productivity of water. *Water Resources Development* 15 (1/2): 55–71.

Morton, E. 2008. The challenge of reducing consumptive use in the Murray-Darling Basin. *Regional Water Conference* Lake Hume Resort.

NCC (National Competition Council). 2004. *Assessment of Governments' Progress in Implementing the National Competition Policy and Related Reforms,* Volume Two: Water, Canberra, National Competition Council.

NWC (National Water Commission). 2008. *Update on Progress of Reform—Input into the Water Sub Group Stocktake Report.* Canberra: National Water Commission.

Productivity Commission. 2006. *Rural Water Use and the Environment: The Role of Market Mechanisms, Research Report.* Melbourne: Productivity Commission.

Randall, A. 1981. Property entitlements and pricing policies for a mature water economy. *Australian Journal of Agricultural Economics* 25 (3): 195–220.

Ribaudo, M. O., D. Colacicco, L. L. Langner, S. Piper, and G. D. Schaible. 1990. *National Resource and Users Benefit from the Conservation Reserve Program.* Agricultural Report 627. Washington: Economic Research Service, Department of Agriculture.

Richter, L.K. 1983. Tourism politics and political science: A case of not so benign neglect. *Annals of Tourism Research* 10 (3): 313–335.

Smith, D. 1998. *Water in Australia: Resources and Management.* Melbourne: Oxford University Press.

Smith, S. 1994. The tourism product. *Annals of Tourism Research* 21 (3): 582–595.

Stonich, S. 1998. The political ecology of tourism. *Annals of Tourism Research* 25 (1): 25–54.

Ward, F., and A. Michelsen. 2002. The economic value of water in agriculture: Concepts and policy applications. *Water Policy* 4: 423–446.

Westernport Water. 2009. *State Government Report Puts the "Oughta" into Water,* accessed October 10, 2009, from www.westernportwater.com.au/News/Details/?NewsID=145.

Wong, P. 2008. $8.6 million for research on 'win-win' water use. Media Release for the Minister for Climate Change and Water, Canberra.

Young, M. 2009. The effects of water markets, water institutions and prices on the adoption of irrigation technology. In *The Management of Water Quality and Irrigation Technologies.* pp. 227–249. Edited by J. Albiac and A. Dinar. London: Earthscan.

CHAPTER 2

The Environmental Status of Australia's Rivers: A Production Systems Perspective

Simon Hone

*I*n the last 10 years, severe droughts throughout much of Australia have substantially reduced the volume of water available for environmental and other uses, while water management has become an increasingly controversial public policy issue. The Wentworth Group of Concerned Scientists (2010) argue that about 30% of average annual diversions in the Murray-Darling Basin should be returned to rivers (in addition to the 10% already obtained from water buybacks and infrastructure projects). This has been opposed by some rural groups, including the National Farmers' Federation, which contends that "the Wentworth Group's suggestions would kill off Australia's food bowl [and] it appears the Wentworth Group, seemingly blinded by dogma, can't see the forest for the trees" (NFF 2010, 1).

The focus in this chapter is on the ecological status of Australia's rivers, in particular, the effects of human intervention on river health. The scientific issues discussed here are central to understanding the policy debate, and this discussion sets the scene for the coming chapters of this book. This chapter examines changes in the health of Australia's rivers and floodplains, covering the development of water resources as well as the effects on hydrology, water quality, and habitat. A simplified representation of the environmental production system is used to conceptualize some of the linkages between inputs and outputs. There are two motives for examining these relationships: first, to explain why many Australian rivers and floodplains have been degraded, and second, to provide evidence on how environmental outputs (such as native fish numbers) might respond to policy interventions (such as buying water for environmental uses).

Accordingly, the chapter is divided into two sections. The first section outlines the environmental status of Australia's river systems; the second examines the reasons for their deterioration and suggests a framework for conceptualizing the relationships between "inputs" (such as surface-water diversions) and environmental "outputs" (such as biodiversity benefits).

CURRENT STATE OF AUSTRALIA'S RIVER SYSTEMS

Unlike national income measures, such as gross national product and gross domestic product, there are no quantitative measures of the overall condition of Australia's rivers, wetlands, and lakes. According to the *State of the Environment* report, "Whether river health has been getting better or worse or has been stable at a national scale over the last five years is difficult to assess because of a lack of data" (ASEC 2006, 62). The absence of reliable data is understandable. In general, it is considerably more expensive to measure the environmental output of a wetland (for example, bird breeding or improved water quality) than that of a manufacturing plant, and in the absence of market prices, there is limited basis for reliably "weighting" different environmental outputs. Although no national measures of river health exist, a number of basin-level assessments have been done. In 2005, the Victorian government released an assessment of trends in river health—the first of its kind in Australia. In 2008, the Murray-Darling Basin Commission released its *Sustainable River Audit* (Davies et al. 2008), which reported on ecological health across the basin. These assessments of river health are discussed in Box 2.1.

Box 2.1. *Evaluations of Ecosystem Health at the River Valley Level*

The Victorian government assessment of trends in river health was based on surveys in 1999 and 2004. In both years, about 1,000 reaches were surveyed across the state, with sample sites being randomly selected. In 2004, data were collected on hydrology (low flows, high flows, zero flows, seasonality, and variability), water quality (phosphorus, turbidity, salinity, and pH), streamside zone (width, longitudinal continuity, understory diversity, and so on), physical form (bank stability, large wood, and fish passage), and aquatic life (macroinvertebrates).

The Victorian Department of Sustainability and Environment (2005) found that 32% of Victorian river length was in "poor" or "very poor" condition, 47% was in "moderate" condition, and 21% was in "good" or "excellent" condition. River basins in eastern Victoria were generally in better condition than river basins in western Victoria. Between 1999 and 2004, there were no substantial changes in the condition of Victoria's major rivers and tributaries. Unfortunately, changes in sampling location and assessment methodology make it difficult to reliably estimate trends for individual reaches and streams.

Between 2004 and 2007, the Murray-Darling Basin Commission undertook a similar assessment of river health in the Murray-Darling Basin. Davies et al. (2008) looked at three themes: hydrology (high flows, low flows, seasonality, variability, and annual volume), fish (expectedness, nativeness, number of species, and biomass), and macroinvertebrates (expectedness, sensitivity to disturbance, and number of families). These themes were chosen partly because of their importance in river ecosystems and sensitivity to human intervention. A condition index was estimated for fish and macroinvertebrate themes for each river valley, taking a value between 0 and 100, with 100 being the estimated condition without "significant human intervention," taking drought into account. The hydrology theme was assessed semiquantitatively and given less weight in determining overall river health.

According to Davies et al. (2008), one valley was rated in "good" health, two valleys were rated in "moderate" health, seven valleys were rated in "poor" health, and 13 valleys were rated in "very poor" health. No valleys were rated in "extremely poor" health. Northern valleys were generally in better health than southern valleys, but substantial variability also occurred within valleys. For example, in the Murrumbidgee Valley in southern New South Wales, the fish condition index was 0 in the upland zone and 42 in the lowland zone. A discrepancy was also evident in the macroinvertebrate condition index, which was 40 in the upland zone and 62 in the lowland zone. Across the Murray-Darling Basin, upland and montane zones tended to be in worse condition than slopes and lowland zones, often as a result of introduced fish.

Many case studies have analyzed the environmental condition of specific areas (for example, the Coorong and Lower Lakes) or a single dimension of environmental output (such as the abundance of frogs). It is difficult to make inferences about the overall condition of Australia's rivers, wetlands, and lakes from these examples. First, they were selected to illustrate a variety of impacts, not selected randomly from the literature. Second, the literature could have a tendency toward analyzing environmentally degraded systems, to the extent that these systems are considered more important, while publication bias is also possible. Case studies examining changes in floodplain habitat, animal populations, and cyanobacterial blooms are discussed below.

Floodplain Vegetation

There is evidence of widespread deterioration in the condition of many of Australia's wetlands. According to Arthington and Pusey (2003), about 90% of floodplain wetlands in the Murray-Darling Basin no longer exist. In New South Wales, about 50% of coastal wetlands have been lost, and about 75% of wetlands on the Swan Coastal Plain in Western Australia have also been lost. Moreover, about 35% of Australian wetlands listed under the Ramsar Convention "have changed in ecological character or have the potential to change" (ASEC 2006, 64).

Kingsford (2000) examined changes in flooding patterns of four wetlands in the Murray-Darling Basin and found substantial changes compared with natural conditions. The Barmah-Millewa Forest covers approximately 65,000 hectares along the River Murray. It is located on the border of Victoria and New South Wales and is notable for having the largest area of river red gums in Australia. The frequency of flooding has declined from 80% to 35% of years under natural conditions. The Chowilla floodplain is located farther downstream in South Australia. Water reaches the floodplain about half as often as under natural conditions—every 2.5 years instead of every 1.2 years—and the area inundated every 10 years has declined from 77% to 54%.

The Gwydir wetlands are terminal wetlands on the Gwydir River in northern New South Wales. The main wetland area was flooded about 17% of the time under natural conditions but is now flooded about 5% of the time. Southwest of the Gwydir wetlands, on the Macquarie River, the Macquarie Marshes cover approximately 130,000 hectares during large floods. The marshes have the largest area of reed beds and red gums in northern New South Wales, and 72 species of waterbirds have been recorded in the marshes. Since the initial development of water resources in the area, the size of the Macquarie Marshes has halved (Kingsford 2000).

As flooding patterns have changed, changes have also occurred in riparian vegetation. For example, in the Barmah-Millewa Forest, the composition of the vegetation community has been altered. In some areas, plants that depend on regular flooding, such as moira grass, have been replaced by red gums, while red gums have been replaced by black box on the margins. Black box trees have been killed on the Chowilla floodplain by saline groundwater discharge and water stress, in some cases after 35 years without flooding. In the Gwydir wetlands, the area covered by marsh club rush has fallen from 2,200 to 700 hectares, and other aquatic vegetation has been replaced by terrestrial vegetation (Kingsford 2000). The Macquarie Marshes have also undergone substantial changes. The area of river red gums halved between 1934 and 1981, and the area of reed beds halved between 1963 and 1972. Several hundred hectares of coolibah trees have died since the 1970s (Bunn and Arthington 2002). Nationally, the length of river covered by riparian vegetation saw a small increase between 1991 and 2004 (ASEC 2006).

The natural flooding of wetlands relies on overbank flows. In some areas, this has been compromised by the erosion of river channels and a reduction in the frequency of high-flow events. The latter has two main causes. First, water diversions and evaporation from dams have reduced the overall volume of water available for the natural environment, as discussed below. Second, dams sometimes capture the flood pulse, thus reducing the variability of streamflow. Whereas some wetlands have received less water, the reduction in the variance of streamflow has caused the permanent inundation of some previously intermittently flooded wetlands. In the Murray and Lower Darling, permanent inundation is the main cause of wetland damage (Norris et al. 2001). About 37,000 hectares of previously intermittently flooded wetlands along the River Murray between Hume Dam and Wellington are now permanently inundated (NSW DPI 2010).

Animals

Extensive research has been conducted on animal species that live in Australia's rivers and wetlands. In the Murray-Darling Basin, native fish populations have declined, with almost half of native fish species being recorded as threatened (Gehrke et al. 2003). A New South Wales survey in 1997 found that native fish comprised only 20% of the survey catch from regulated rivers in the Murray region. Moreover, no Murray cod or freshwater catfish were caught, despite intensive fishing over two years at 20 randomly chosen sites. These species, and native fish species in general, were more abundant in the Darling region (Harris and Gehrke 1997). Similarly, Davies et al. found native fish in only 43% of Murray-Darling Basin valleys where they were predicted, noting that this confirmed the "well known decline in native fish in the Basin" (2008, 62).

At the same time, the distribution and abundance of alien species, such as common carp, gambusia, and goldfish, have substantially increased. Common carp was the dominant species in the *Sustainable River Audit* surveys, accounting for 58% of biomass (Davies et al. 2008). To take an extreme example of carp infestation, in the abovementioned New South Wales survey, one common carp was found per square meter of river surface area in a lowland reach of the Bogan River (Harris and Gehrke 1997).

Bird populations have also declined in some areas. In the Barmah-Millewa Forest, a number of bird species have stopped breeding in the forest, including brolgas, glossy ibis, little egrets, and whiskered terns. Some other species still breed, but in reduced numbers (Kingsford 2000). In the northern part of the Macquarie Marshes, the number of species decreased between 1983 and 1993 (Kingsford and Thomas 1995).

Frogs and some species of invertebrates are often considered to be sensitive toward changes in the health of aquatic ecosystems. Across Australia, 27 species of frogs are listed as endangered or vulnerable. Since European settlement, about one-third of Australia's rivers (by length) have lost at least 20% of aquatic invertebrate species. Between 1990 and 2004, the Australian River Assessment System was used to examine the condition of macroinvertebrate communities at some 4,700 sites covering all Australian states and territories. About 45% of sites were found to be "impaired" relative to the reference condition, which refers to the predicted condition in the absence of environmental stress, such as pollution or habitat degradation (ASEC 2006).

According to Gehrke et al., "Macroinvertebrate communities have been altered substantially in many areas [of the Murray-Darling Basin], with the disappearance of Murray crayfish, and some species of freshwater snails and mussels from reaches in South Australia" (2003, 2). This is also evident in the Barmah-Millewa Forest, where, in the 1930s, more than 25,000 leeches a year were gathered for use in hospitals. Leeches were seldom seen after the 1970s (Kingsford 2000). In the Lower River Murray, the populations of many species of snails have declined markedly over the last 50 years, with only 1 of 18 species remaining common. *Notopala sublineata* seems to be locally extinct, and 2 other species survive only in irrigation pipelines (Sheldon and Walker 1997).

Cyanobacterial Blooms

Blooms of cyanobacteria, or blue-green algae, occur naturally in many Australian rivers and lakes. In 1830, the explorer Charles Sturt observed cyanobacterial blooms on the Darling River, and in 1878, cyanobacterial blooms in Lake Alexandrina caused animal deaths (MDBA 2010a). Cyanobacterial blooms can appear rapidly under favorable conditions and are often typified by green scum on the water surface. In addition to threatening some native animals and plants, these blooms pose a health risk to people, causing gastroenteritis and skin irritations (Bowling and Baker 1996). Intense cyanobacterial blooms are becoming more likely, according to the Murray-Darling Basin Authority (2010a), with major blooms throughout 800 kilometers of the River Murray in 1983 and 2009. The largest bloom occurred in 1991, at its peak covering about 1,000 kilometers of the Darling River. During this outbreak, Bowling and Baker (1996) found high concentrations of the cyanobacteria *Anabaena circinalis* at many locations, which was determined to be toxic in laboratory tests and implicated in stock deaths.

REASONS FOR THESE ENVIRONMENTAL CHANGES

The above examples show that many of Australia's rivers, wetlands, and lakes have deteriorated over time. Why have these changes happened? And more specifically, what has been the impact of human intervention? This chapter adopts a production systems perspective and follows Poff et al. (1997) in identifying hydrology, water quality, and physical habitat as fundamental components of a river ecosystem (Figure 2.1).

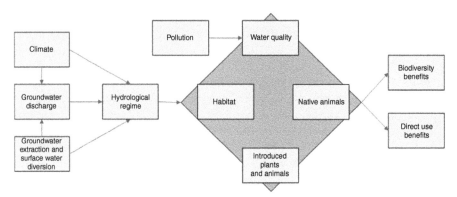

Figure 2.1 *A simplified environmental production system showing key relationships between environmental inputs and outputs*

In many areas, the development of water resources has had a substantial impact on these components. Australia is a dry continent, on average, with lower rainfall and runoff per square kilometer than any other continent except Antarctica (Arthington and Pusey 2003). Although 80% of Australia experiences less than 600 millimeters of rainfall a year, coastal Australia is generally wetter, and some parts of the North Queensland coast average over 4,000 millimeters of rainfall a year. Australia also has higher riverflow variability than any other continent (Kirby et al. 2006). Indeed, Cooper Creek and the Diamantina River, in central Australia, have the highest flow variability in the world according to some measures (Arthington and Pusey 2003).

In 1886, rainfall variability motivated the Victorian government and brothers William and George Chaffey, Canadian-born engineers and irrigation planners who had developed irrigation colonies in California, to develop Australia's first irrigation settlement at Mildura on the River Murray. Since the 1800s, thousands of flow regulation structures have been built throughout Australia, including about 450 large dams and 50 substantial water transfer schemes, and the nation now has the highest per capita water storage capacity in the world. More development has occurred in southern Australia than in the northern part of the country, with rivers in the tropics and subtropics largely unmodified apart from coastal Queensland and the Ord River in Western Australia (ASEC 2006). The Murray-Darling Basin has some 3,600 weirs as well as other structures, such as locks and floodplain levee banks (Arthington and Pusey 2003).

The construction of public and private storages increased the reliability of water supply, allowing the development of a large irrigation industry. In the Murray-Darling Basin, diversions, mainly for irrigation, increased gradually until the 1990s (Figure 2.2). In response to this trend, the Murray-Darling Basin Ministerial Council broadly agreed to limit overall diversions to 1994 levels, adjusted for

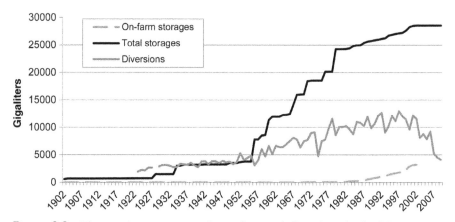

Figure 2.2 *Changes in storage capacity and annual diversions in the Murray-Darling Basin, 1902–2009*

Source: Productivity Commission (2010)

seasonal climatic conditions (Productivity Commission 2010). An extended drought during the 2000s resulted in a marked decline in diversions.

As surface water has become more expensive, because of drought and limits on overall diversions, many farmers have begun to substitute it with groundwater. In the Lower Murrumbidgee, groundwater use increased from about 100 to 300 gigaliters between 1994 and 2002 (Goesch et al. 2007). Nationally, an 88% increase in groundwater use was seen between 1984 and 1997 (ASEC 2006).

The link between groundwater and surface water can be complex. For example, groundwater can be recharged by rivers (known as "losing" rivers), while groundwater can also contribute to rivers ("gaining" rivers). Connectivity can be complex, with a single river potentially both gaining and losing water, depending on location and time (Goesch et al. 2007). The proportion of water in unregulated rivers that comes from groundwater can sometimes exceed 75% (Kirby et al. 2006). Under these conditions, groundwater extractions can have a substantial impact on surface-water availability. Moreover, there will sometimes be lags. In large regional groundwater systems, such as the Great Artesian Basin in central Australia, it can take many thousands of years for water to move through the system. By contrast, small and localized individual fractured rock aquifers cover most of the eastern Murray-Darling Basin, and these are considerably more responsive, on a scale of decades, to changes in water availability (Kirby et al. 2006).

As discussed above, new rules were introduced in the 1990s to limit the decline in the volume of water allocated to environmental uses in the Murray-Darling Basin. In the 2000s, Australian governments sought to substantially increase environmental allocations through programs such as the A\$700 million Living Murray Initiative, A\$3.1 billion Restoring the Balance Program, and A\$5.8 billion Sustainable Rural Water Use and Infrastructure Program.[1] As of early 2010, about 460 gigaliters of water had been recovered through water buybacks and infrastructure under the Living Murray Initiative, and 530 gigaliters under Restoring the Balance water buybacks (Productivity Commission 2010). Smaller projects have also been undertaken, such as the one discussed in Box 2.2.

Box 2.2. *Restoration of the Snowy River*

The Snowy Mountains Hydro-Electric Scheme was completed in 1974, diverting approximately 1,000 gigaliters annually from the Snowy River. This is equivalent to about 99% of mean natural flow at Jindabyne. Understandably, this has resulted in various environmental problems, which New South Wales and Victorian governments hope to partially address through returning up to 21% of mean natural flow. Thus far, about 145 gigaliters of entitlements have been recovered for the Snowy River (NSW Office of Water, 2010).

Impacts of Development

This section examines the effects of development on key components of the environmental production system: hydrology, water quality, and habitat.

Hydrology.   The development of water resources has changed the hydrology of many Australian rivers. Poff et al. (1997) discuss the following key dimensions of the flow regime:

- *Magnitude*: the volume of water that moves past a location over a given time interval. In Figure 2.3, the magnitude of flows between 2005 and 2010 is given by the shaded area under the curve.
- *Frequency*: how often flow exceeds a given value over a time interval. In Figure 2.3, flow exceeded 100 megaliters per day on two occasions in five years.
- *Duration*: the period of time associated with a specific "flow condition." In Figure 2.3, flows in excess of 100 megaliters per day were maintained for about two months on both occasions.
- *Seasonality, or timing*: the regularity of a specific "flow condition." In Figure 2.3, annual peaks occur with high seasonal predictability (in late spring each year).
- *Rate of change*: how quickly flow changes over time. In Figure 2.3, this is given by the slope of the line, with a steeper line indicating a faster rate of change.

According to Arthington and Pusey (2003), these dimensions of hydrology have been fundamentally altered in many of Australia's great rivers, including the River Murray, Darling River, coastal rivers in New South Wales and Queensland, the Gordon River in Tasmania, and the Ord River in Western Australia. The overall magnitude of flows has been markedly reduced in many rivers and streams, while

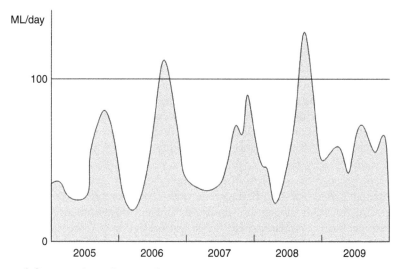

Figure 2.3 *Hypothetical streamflow over time*

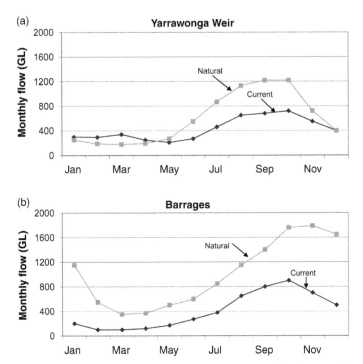

Figure 2.4 *Simulated flows under "natural" and "current development" conditions at the Yarrawonga Weir (a) and the Barrages (b)*

Source: Productivity Commission (2010)

seasonal flow patterns have been lost or inverted. These trends are evident in simulated "natural" and "current development" flows at the Yarrawonga Weir and Barrages on the River Murray (Figure 2.4). (These simulations assume an identical climate to better isolate the effects of the development of water resources on average monthly flow.) The upstream location, Yarrawonga Weir, has experienced a 30% decrease in simulated mean annual flow. At the Barrages, the decline in simulated mean annual flow is substantially larger, at about 60%. Natural seasonal flow patterns have been severely disrupted at Yarrawonga Weir, with a substantial decrease in flow between July and October and a small increase between January and April, the latter due mainly to irrigation water deliveries (Productivity Commission 2010).

The measures discussed above are monthly averages and do not provide information on the variability of flow between years (across states of nature). This is relevant because the proportional difference between "natural" and "current development" flows at the Barrages increases considerably in dry years. For example, at the 90th percentile, simulated "current development" flow is 360 gigaliters per year, compared with 5,200 gigaliters under "natural" conditions (Kingsford et al. 2009).

Poff et al. note that the flow regime is a "master variable" that "limits the distribution and abundance of riverine species and regulates the ecological integrity of flowing water systems" (1997 769). As well as directly influencing ecological integrity, the flow regime has indirect effects through water quality, physical habitat, energy sources, and biotic interactions. For example, additional streamflow can dilute salt loads, reducing salinity.

Water Quality. Many Australian rivers are naturally saline; however, salinity has increased in some areas as a result of irrigation and land clearing. In 2001, about one-third of catchments assessed by the *National Land and Water Resources Audit* exceeded their salinity thresholds. These catchments were mainly in the southern Murray-Darling Basin, on the southeast coast, and in southwest Western Australia. In the Murray-Darling Basin, four "salinity-affected" catchments became less saline over the 1990s, while three affected catchments became more saline (ANRA 2000). There is no evidence of general increases in salinity in the northern Murray-Darling Basin (Kirby et al. 2006).

Australia's agricultural soils often contain substantial quantities of salt. In parts of Western Australia, salt may have been carried by winds from the ocean, whereas in other areas, salt may originate from the erosion of parent rock or from ancient drainage basins and inland seas. Irrigation and land clearing contribute to river salinity. These activities raise the groundwater table—irrigation by adding more water, and land clearing by reducing the amount of surface water and groundwater used by vegetation (native vegetation tends to intercept and use more water than do annual crops and pastures). The rising groundwater table dissolves salt lying dormant in the soil profile. Eventually, this saline groundwater may evaporate on the surface or discharge into surrounding rivers (AAS 2001; Pakula 2004).

The impact of irrigation and land clearing at a given location depends on factors such as the gradient of the groundwater system, which determines the direction and speed of groundwater movement. For example, there may be negligible effect if the gradient is away from the river; when the gradient is toward the river, the timing of impacts will be influenced by the gradient and the distance from the river (Pakula 2004). This can lead to extended time lags. In the Riverland area of South Australia, the lags between changing land use and the resulting impacts on the river can be more than 50 years (Barnett and Yan 2004).

The impact of irrigation also depends on the extent to which the salt in irrigation water does not return to the river (that is, the proportion of salt imports retained in the system). Salt can be trapped in soil, shallow groundwater, and evaporation basins or diverted to wetlands. In the Kerang region, unusually little salt is retained because the irrigation district was built on a natural groundwater discharge area. As a consequence, salt exports are about six times greater than salt imports (Kirby et al. 2006). Salt loads can also be influenced by withdrawing water from the natural environment for irrigation, which tends to reduce flow and, among other impacts, increases salinity for a given salt load.

Whereas irrigation and land clearing have increased salinity in some areas, other human activities, such as the development of salt interception schemes and changing land management practices, have had the opposite effect. According to the Murray-Darling Basin Ministerial Council (2008), river salinity at Morgan in

2006–2007 would have been 200 to 350 electrical conductivity units higher without these interventions.

The *National Land and Water Resources Audit* (ANRA 2000) also looked at three other dimensions of water quality: nutrients, turbidity, and pH. Nutrients were a "major water quality issue" in 61% of assessed catchments, with nutrient levels generally higher in more developed areas. Like salinity, nutrient levels were increasing in some affected catchments and decreasing in others. During the cyanobacterial bloom on the Darling River in 1991, total phosphorus levels were very high, substantially exceeding Australian and New Zealand Environment Conservation Council guidelines for "good water quality" at many sites (Bowling and Baker 1996). Turbidity was a "major water quality issue" in 61% of assessed catchments, and pH in 16%. Turbidity has been increasing in many Australian rivers, while there were trends toward increasing acidity in some catchments (particularly inland Victoria) and increasing alkalinity in others (ANRA 2000).

These changes in water quality are likely to be caused at least partly by human intervention. According to the Australian State of the Environment Committee, "For most rivers, the largest source of increased phosphorous loads is gully and stream bank erosion [while] increased nitrogen levels are typically from fertilizer use, animal waste and sewage discharges" (2006, 65). Gully and streambank erosion caused by livestock and land clearing can also result in an increase in turbidity (MDBC n.d.), and increased acidity in some Victorian rivers could be due to land degradation (ANRA 2000).

Physical Habitat. Changes in hydrology have also contributed to changes in habitat within rivers and on floodplains. According to Poff et al. (1997), the physical habitat of a river includes sediment size and heterogeneity, the shape of the river channel, and other geomorphic features. This physical habitat is determined largely by physical processes, especially the movement of water and sediment within the river channel and between the river channel and floodplain. Dams capture all but the finest sediment moving downstream. In Australia, more than 20 public reservoirs became fully silted between 1890 and 1960 (Chanson 1998). Without these sediments, water releases can erode smaller sediments from the downstream channel, leading to coarsening of the riverbed. Moreover, erosion of the channel and "tributary headcutting" can occur. Dams and diversions can also reduce the magnitude and frequency of high-flow events. In their literature review, Poff et al. explain that this can cause a number of geomorphic responses, including "deposition of fines in gravel; channel stabilization and narrowing; and reduced formation of point bars, secondary channels, oxbows and changes in the channel planform" (1997, 773).

Changes in hydrology have other impacts on habitat. In the Goulburn River below Lake Eildon, riffle habitat is important for many species of invertebrates and fish. Riffle habitat is a shallow, fast-flowing, and turbulent zone. An "inverted U" type relationship exists between releases from Lake Eildon and riffle habitat downstream. At reach one, releases of 10 megaliters per day are associated with a riffle area of 4.5 square meters per meter (m^2/m), releases of 100 megaliters per day with a riffle area of 6.5 m^2/m, and releases of 10,000 megaliters per day with a riffle area of about 1.5 m^2/m. Because of irrigation deliveries, current mean daily summer flow is approximately 9,000 megaliters per day, which is substantially higher than under natural conditions. This change in hydrology has roughly halved the area of riffle habitat between December and April (Cottingham et al. 2003).

Aside from hydrology, human intervention has affected river habitats by introducing barriers to fish movement, such as weirs, and clearing snags, and the development of river catchments has increased sediment loads in rivers and estuaries. According to the *State of the Environment* report (ASEC 2006), large volumes of sediment have been deposited in many Australian river channels, especially in low-gradient areas. Large deposits of sand and other fine materials have smothered instream habitats in lowland reaches, and rehabilitation techniques are currently being explored.

The Environmental Production System

So far, this chapter has examined changes in the health of Australia's rivers and floodplains, as well as the development of water resources and its effects on hydrology, water quality, and habitat. In an effort to complete the explanation of why many Australian rivers and floodplains have been degraded and provide evidence on how environmental outputs might respond to policy interventions, Figure 2.1 shows some of the linkages between inputs and outputs. This is a simplified representation of the environmental production system.

The degradation of some Australian river and wetland ecosystems—a change in outputs—has occurred alongside the development of those water resources for irrigation and other uses, as well as changes in land use. At a conceptual level, there is good reason to infer some overall causal relationship between human development and ecosystem health. Plants and animals have evolved to exploit certain habitats and conditions. As discussed above, human intervention has altered many of these habitats, often reducing variability that was a natural part of most Australian river systems.

This variability is central. According to Poff et al., "Over periods of years to decades, a single river can consistently provide ephemeral, seasonal, and persistent types of habitat that range from free-flowing, to standing, to no water. This predictable diversity of in-channel and floodplain habitat types has promoted the evolution of species that exploit the habitat mosaic created and maintained by hydrologic variability. For many riverine species, completion of the life cycle requires an array of different habitat types, whose availability over time is regulated by the flow regime" (1997, 772). Variability also contributes to overall biological diversity and ecosystem function, as some species do best in wet conditions and others in dry conditions.

This intuition is backed by specific studies. For example, Poff et al. (1997, 776) have associated flow stabilization and loss of seasonal flow peaks with the following impacts:

• altered energy flow;
• invasion of exotic species;

- seedling desiccation and ineffective seed dispersal;
- disrupted cues for fish spawning, hatching, and migration;
- loss of fish access to wetlands and backwaters;
- modification of the aquatic food web structure; and
- reduced plant growth rates.

This is not to suggest that all changes in the health of Australia's rivers and wetlands can be attributed to human development. Some parts of Australia experienced prolonged drought during the 2000s. Inflows to the Murray-Darling Basin averaged about 2,000 gigaliters a year between 2006 and 2009, whereas the long-term average is 11,000 gigaliters a year (MDBA 2010b). Drought most likely has contributed to some of the adverse results discussed above.

Figure 2.1 also reveals that, although important, hydrology is not the only variable that influences ecosystem health. For example, the introduction of exotic species and changes in water quality can affect native plants and animals independently of hydrology. This is evident in the Kiewa Valley in northeast Victoria, where native fish communities are in poor condition, despite hydrology being largely unaffected by human intervention (Davies et al. 2008).

The relationship between inputs and environmental outputs is usually characterized by complexities, such as the following:

- *Lagged effects.* These are central to understanding the relationship between groundwater and surface water. According to Evans, "Between the start of pumping and an impact in the stream, the lag can be hours, weeks, years, or even centuries. When pumping ceases, it may take decades before stream flow returns to its previous norm" (2007, 7). Kingsford (2000) discusses the lagged effects of changes in hydrology on ecological systems.
- *Irreversibility.* A change in inputs could also lead to an irreversible change in environmental outputs. In the Murray-Darling Basin, 35 bird species and 16 mammal species are endangered (DEWHA 2010). While their contributions to ecological systems and human well-being are relevant and debatable issues, their extinction could not be reversed with existing technology.
- *Nonlinearities.* Many geomorphic and ecological processes show nonlinear responses to flow. As expressed by Poff et al., "Clearly, half of the peak discharge will not move half of the sediment, half of the migration motivational flow will not motivate half of the fish, and half of an overbank flow will not inundate half of the floodplain" (1997, 781).
- *Interactions.* The environmental output of a wetland might depend on inputs applied elsewhere, including to other parts of the river system and floodplains. According to Environment Victoria, "There is a tendency to see rivers as a series of disconnected assets or drought refuges [but] if a river system is to survive and thrive, it is essential that it retains both lateral and longitudinal connectivity. In other words, it needs enough water for fish and other animals to migrate along it, and to retain connection to its floodplains, which serve as the larders of the river system" (Productivity Commission 2010, 74).

These complexities make it difficult to quantify the environmental relationships summarized in Figure 2.1. Furthermore, in the absence of experiments, variables that affect the health of river systems, such as water diversions and changes in land use, sometimes move together. Exacerbated by lags, this makes it difficult to separate the influence of different variables. In writing their review, Bunn and Arthington "often encountered reports of river systems affected by multiple stressors and [were] unable to definitively separate the impacts of altered flow regimes from those of the myriad of other factors and interactions." They concluded that "our limited ability to predict and quantify the biotic response to flow regulation is a major constraint" to achieving better management (2002, 505).

Clearly, understanding of the complex relationships that exist within river ecosystems is far from complete, and this gap serves to confound the policy arena that seeks to secure improved environmental outcomes for this region. For tourism and recreation, both of which rely heavily on the various aspects of amenity delivered by rivers and other bodies of water, this lack of understanding translates to a fraught policy context. The remaining chapters of this book address many of the issues and challenges of the tourism and recreation sector in an effort to gain understanding of the complicated nexus among water, tourism, and recreation.

NOTE

1. A$1 = US$0.9978 as of January 2011.

REFERENCES

AAS (Australian Academy of Science). 2010. *Salinity: The Awakening Monster from the Deep*, accessed September 29, 2010, from www.science.org.au.

ANRA (Australian Natural Resource Atlas). 2000. *Water Quality*, accessed September 29, 2010, from www.anra.gov.au.

Arthington, A.H., and B.J. Pusey, 2003. Flow restoration and protection in Australian rivers. *River Research and Applications* 19: 377–395.

ASEC (Australian State of the Environment Committee). 2006. *Australia State of the Environment 2006*. Canberra: Department of the Environment and Heritage.

Barnett, S.R., and W. Yan. 2004. *Groundwater Modelling of Salinity Impacts on the River Murray due to Vegetation Clearance in the Riverland Area of South Australia*. Adelaide: Department of Water, Land and Biodiversity Conservation.

Bowling, L., and P. Baker. 1996. Major cyanobacterial bloom in the Barwon–Darling River, Australia, in 1991, and underlying limnological conditions. *Marine and Freshwater Research* 47 (4): 643–657.

Bunn, S.E., and A.H. Arthington. 2002. Basic principles and ecological consequences of altered flow regimes for aquatic biodiversity. *Environmental Management* 30: 492–507.

Chanson, H. 1998. Extreme reservoir sedimentation in Australia: A review. *International Journal of Sediment Research* 13 (3): 55–63.

Cottingham, P., M. Stewardson, D. Crook, T. Hillman, J. Roberts, and I. Rutherfurd. 2003. *Environmental Flow Recommendations for the Goulburn River below Lake Eildon*. Canberra: CRC Freshwater Ecology and CRC Catchment Hydrology.

Davies, P.E., J.H. Harris, T.J. Hillman, and K.F. Walker. 2008. *Sustainable Rivers Audit Report 1: A Report on the Ecological Health of Rivers in the Murray-Darling Basin, 2004–07.* Canberra: Murray-Darling Basin Commission.

DEWHA (Department of Environment, Water, Heritage and the Arts). 2010. *Murray-Darling Basin,* accessed September 29, 2010, from www.environment.gov.au.

Evans, R. 2007. *The Impact of Groundwater Use on Australia's Rivers: Exploring the Technical, Management and Policy Challenges.* Canberra: Land and Water Australia.

Gehrke, P., B. Gawne, and P. Cullen. 2003. *What Is the Status of River Health in the MDB?* accessed September 29, 2010, from www.clw.csiro.au.

Goesch, T., S. Hone, and P. Gooday. 2007. *Groundwater Management: Efficiency and Sustainability.* Canberra: Australian Bureau of Agricultural and Resource Economics.

Harris, J.H., and P. Gehrke. 1997. *Fish and Rivers in Stress: The NSW Rivers Survey.* Cronulla and Canberra: NSW Fisheries Office of Conservation and Cooperative Research Centre for Freshwater Ecology.

Kingsford, R.T. 2000. Ecological impacts of dams, water diversions and river management on floodplain wetlands in Australia. *Austral Ecology* 25: 109–127.

Kingsford, R., P.G. Fairweather, M.C. Geddes, R.E. Lester, J. Sammut, and K.F. Walker. 2009. *Engineering a Crisis in a Ramsar Wetland: The Coorong, Lower Lakes and Murray Mouth, Australia.* Sydney: Australian Wetlands and Rivers Centre, University of New South Wales.

Kingsford, R.T., and R.F. Thomas. 1995. The Macquarie Marshes in arid Australia and their waterbirds: A 50-year history of decline. *Environmental Management* 14: 867–878.

Kirby, M., R. Evans, G. Walker, R. Cresswell, J. Coram, S. Kahn, Z. Paydar, M. Mainuddin, N. McKenzie, and S. Ryan. 2006. *The Shared Water Resources of the Murray-Darling Basin.* Canberra: Murray-Darling Basin Commission.

MDBA (Murray-Darling Basin Authority). 2010a. *Blue-Green Algae in the River Murray,* accessed September 29, 2010, from www.mdba.gov.au.

MDBA (Murray-Darling Basin Authority). 2010b. *Annual Report,* accessed September 29, 2010, from www.mdba.gov.au.

MDBC (Murray-Darling Basin Commission). no date. *Water Quality,* accessed September 29, 2010, from www.mdba.gov.au.

MDBMC (Murray-Darling Basin Ministerial Council). 2008. *Report of the Independent Audit Group for Salinity 2006–07.* Canberra: Murray-Darling Basin Commission.

NFF (National Farmers' Federation). 2010. *NFF Slams Blinkered Group on Murray-Darling Water Cuts,* accessed September 20, 2010, from www.nff.org.au.

Norris, R.H., P. Liston, N. Davies, J. Coysh, F. Dyer, S. Linke, I.P. Prosser, and W.J. Young. 2001. *Snapshot of the Murray-Darling Basin River Condition.* Canberra: CSIRO Land and Water.

NSW DPI (New South Wales Department of Primary Industries). 2010. *Status of Aquatic Habitats,* accessed September 29, 2010, from www.dpi.nsw.gov.au.

NSW Office of Water. 2010. *Returning Environmental Flows to the Snowy River.* Sydney: NSW Office of Water.

Pakula, B. 2004. *Irrigation and River Salinity in Sunraysia: An Economic Investigation of an Environmental Problem.* Melbourne: Department of Primary Industries.

Poff, N.L., J.D. Allan, M.B. Bain, J.R. Karr, K.L. Prestegaard, B.D. Richter, R.E. Sparks, and J.C. Stromberg. 1997. The natural flow regime. *BioScience* 47 (11): 769–784.

Productivity Commission. 2010. *Market Mechanisms for Recovering Water in the Murray-Darling Basin.* Melbourne: Productivity Commission.

Sheldon, F., and K.F. Walker. 1997. Changes in biofilms, induced by flow regulation, may explain the extinction of gastropods in the Lower River Murray. *Hydrobiologia* 347: 97–108.

Victorian Department of Sustainability and Environment. 2005. *Index of Stream Condition: The Second Benchmark of Victorian River Condition.* Melbourne: Department of Sustainability and Environment.

Wentworth Group. 2010. *Sustainable Diversions in the Murray-Darling Basin: An Analysis of the Options for Achieving a Sustainable Diversion Limit in the Murray-Darling Basin,* accessed September 20, 2010, from www.wentworthgroup.org.

CHAPTER 3

Challenges of Estimating the Value of Tourism and Recreation in the Murray-Darling Basin

Darla Hatton MacDonald, Sorada Tapsuwan, Sabine Albou, and Audrey Rimbaud

T he Murray-Darling Basin provides the backdrop for a growing tourism and recreation sector (Hassall and Gillespie 2004). This sector encompasses a range of activities, from traditional pursuits such as duck hunting, fishing, and boating to, more recently, eco-based tourism and recreation and a flourishing food and wine culture. In combination, tourism, recreation, and amenity values in the Murray-Darling Basin provide jobs and livelihoods for small tourism businesses in the region.

However, the risk of reduced inflows and the current property right arrangements for water pose significant challenges to the health of the sector in this region. Earlier in this book, it has been argued that an understanding of value is a critical precursor to exerting policy influence. It was also noted in Chapter 1 that while this calculation is relatively straightforward in the case of agriculture or industry, consideration of the value of water for tourism and recreation is much more complex. Market values are seldom revealed and are subsequently omitted from consideration in water management decisions.[1]

Worldwide, there has been growing attention to estimating the value of recreation and discussion of how to incorporate such value in water management.[2] Understanding the value of water for tourism, recreation, and amenity is a vital step in the evaluation of current and future water-sharing strategies. In particular, the Water Act 2007 in Australia requires that water be managed to maximize social, economic, and environmental outcomes.

This chapter reviews the existing research on the economic value of tourism and recreation in the Murray-Darling Basin. The remainder of the chapter is divided into four parts. We begin by examining the concept of economic value, then look at methods for estimating quantitative economic values associated with tourism, recreation, amenity, and nonuse values. Following this is an exploration of the idea of total economic value as a means of sorting through the different types of values discussed in the literature, along with a review of published work from the last 30 years. This section is summarized in Table 3.1 (page 40). Finally, we conclude by outlining the gaps in the research undertaken to this point and suggesting some avenues for future research.

THE CONCEPT OF VALUE AND METHODS OF VALUATION

The "value" of tourism and recreation is more than a strict financial aggregation of tourists' expenditures. A relaxing afternoon of fishing would seem to have value. Similarly, time spent in a winery overlooking a wetland has value that exceeds the cost of fuel and the cost of the glass of wine. These values are related and can be portrayed as interlinked circles (Figure 3.1). The nature of the overlap of values can be demonstrated by examples. Past experience with a natural environment is likely to be part of information gathering required to make major purchases such as a holiday property. Potential homeowners are unlikely to buy property near the River Murray unless they have spent time in the area in the past as tourists or engaged in recreational activities. Nonuse values associated with the area may also be associated with past direct experiences such as family camping trips, where a place attachment is retained for the region after moving away. The relative size and overlap of the rings are context-specific.

Value, in its simplest definition, is the quality of something that makes it more or less desirable, useful, estimable, or important to people. Using this definition, tourism and recreation have a number of tangible and intangible benefits that can be summarized in terms of the market-based and nonmarket values associated with the landscape. As such, the value of a red gum forest might be estimated in terms of its extracted timber, but it also provides recreational opportunities for hiking, camping, and bird-watching. Economics has developed a whole set of tools that can elicit and unpack these values, which can be market or nonmarket values.

Market Values

At a national level, gross domestic product (GDP) is the sum of all goods and services made within a country in a given year. The contribution of tourism to

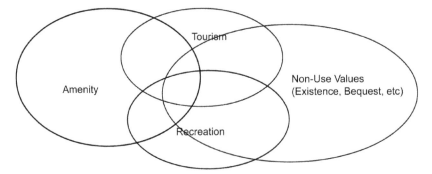

Figure 3.1 *Venn diagram of potential value overlap*

GDP is an aggregation of market values and is published by the Australian Bureau of Statistics (ABS). The ABS estimates that tourism and recreation contributed A\$88.7 billion to GDP in 2007–2008.[3] Tourism is labor-intensive, and 497,800 people in 2007–2008 were directly employed in the Australian tourism industry (ABS 2009).

Computable General Equilibrium Modeling. To examine the impact of changes in policies, economic models such as input–output (IO) and computable general equilibrium (CGE) models have been developed. An example is The Enormous Regional Model (TERM), a multiregional CGE model for Australia developed by the Centre of Policy Studies (CoPS) at Monash University. Its first major application was for a study of the 2002–2003 drought in southeastern Australia (Horridge et al. 2005).

The relevant version of TERM for the tourism industry includes water accounts and subregional data on industries and households in the Murray-Darling Basin. As an example of how the model can be adapted, Young et al. (2006) explored the economic impacts for Australian cities of water supply options like desalination and urban-rural water trading. With minimal population growth, 15% less water in the eastern states due to climate change, and no new water supply options, the scarcity price of water would increase tenfold by 2032. When options, such as urban-rural water trade and desalination, are introduced in the model, only a threefold increase in the price of water would be required for supply to equal demand.

Nonmarket Values

The calculation of GDP or CGE modeling of a regional economy does not include many of the activities that contribute to human well-being. For instance, activities outside the market economy, such as volunteerism, are not included in GDP. Nonmarket values such as the value of ecosystem services of natural capital or the personal health benefits from an afternoon of fishing are similarly not included in GDP. Only the expenditures associated with the activity are included in GDP. The value that people place on the existence of healthy wetland ecosystems and the diversity of bird, plant, and animal species supported within the wetland are not included. Although advances have been made in using "green GDP" (Boyd 2007), these are not widespread or standardized methodologies.

Use Values. Use values arise from the actual use or consumption of the environmental good (Pearce and Moran 1994). Use values can be further separated into direct use values, such as harvesting timber, and indirect use values, such as clean air from trees. Direct use values can be either consumptive, such as harvesting fuel wood, or nonconsumptive, such as visual amenities. Recreational uses are direct use of the natural resource, but unlike timber, their values are not revealed in markets (Costanza et al. 1997).

A nonmarket valuation technique called revealed preference (RP) is commonly used for estimating the use value of environmental goods and services. The RP technique finds the value people place on environmental goods from observed behavior in markets for related goods (Hanley et al. 2001). The two main revealed preference methods are the travel cost method (TCM) and hedonic pricing method (HPM). The TCM values the recreational use of an environmental good, such as a natural park, by estimating the cost individuals incur to travel there. The more people value their experiences at the site, the farther they would travel and the more they would pay to be there.

The HPM uses market prices of other goods to value environmental assets. Property sale prices are commonly used as proxies for estimating the value of environmental goods. The main idea is that the price of a property is affected not only by its structural features (such as bedrooms, bathrooms, and land area), but also by its proximity to environmental locations such as parks and wetlands.

Nonuse Values. Nonuse values are intrinsic values that are not directly related to the present use or consumption of the environmental good, but still have an impact on the well-being of an individual (Nunes 2002). John Krutilla (1967) first introduced the concept of nonuse values into mainstream economic literature in his article "Conservation Reconsidered." Nonuse values include existence, bequest, and option (or quasioption) values. Existence value is the value the public places on knowing something is there, such as knowing that the turtles in wetlands are alive as long as the wetlands do not go dry. Bequest value is the value that people place on being able to conserve the environmental assets for future generations. Goods and services may also be valued for their potential to be available in the future and constitute an option value. It may be thought of as an insurance premium to ensure the supply of the environmental or recreational good later in time. Society may be willing to pay to preserve biodiversity or genetic materials or the potential information associated with the resource to ensure that these options are available in the future. As such, option and quasioption values are both use and nonuse values. Figure 3.2 is a useful summary of how these components fit together and indicative of the care that must be taken with respect to option and quasioption values.

Because of the nonuse nature of some environmental goods and services, their values generally are not revealed in the marketplace. As such, the RP technique is not applicable. The stated preference (SP) technique involves using surveys to elicit people's willingness to pay (WTP) or willingness to accept (WTA) a compensation for a change in environmental quality through a hypothetical market environment. This enables the SP method to capture both use and nonuse values of environmental goods and services. The two main types of SP methods are the contingent valuation method (CVM) and choice modeling (CM), or choice experiments (CE), method.

The CVM involves asking people, through the use of several interview or survey techniques, whether they are willing to pay to prevent a negative change, such as the prevention of deforestation, or for an improvement in, say, the water quality of rivers. Alternatively, people could also be asked whether they are willing to accept compensation for forgoing their rights to an environmental good or service.

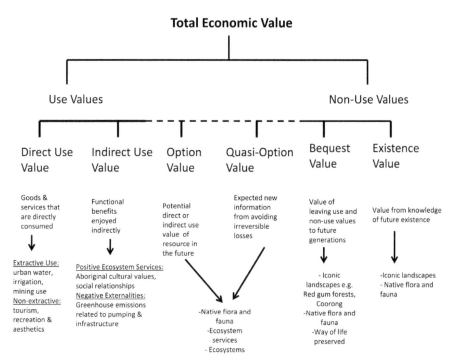

Figure 3.2 *Summary of total value, use value, and nonuse value*

Source: Based on Rolfe (2010)

The CM or CE method is also a survey-based technique for measuring WTP. The environmental good in question is described in terms of its attributes and the levels that the attributes take. The attributes of the River Murray, for example, can include the number and population of fish species, water clarity, and water level. Respondents are asked to rank, rate, or choose their most preferred alternative. By including the price or cost as one of the attributes of the good, WTP can be indirectly recovered from people's rankings, ratings, or choices (Hanley et al. 2001). The difference between the methods is that the CVM can measure only one or two alternatives, whereas the CM or CE method can evaluate several alternatives at one time.

AGGREGATION OF VALUES (TOTAL ECONOMIC VALUE)

For economists, the total economic value (TEV) of an environmental asset comprises use and nonuse values combined. It is basically the net sum of all relevant WTPs for any change in the well-being of people due to changes in the environmental goods and services that have been aggregated across individuals or over time (Bateman et al. 2002). Depending on the type of good and service and data availability, nonmarket valuation techniques are normally used to extract the WTPs for use and nonuse values.

Estimates from nonmarket valuation techniques, such as the elicitation of WTP, are sometimes presented "at the margin." Valuing environmental goods and services at the margin means determining the human welfare change from the relatively small changes in these goods and services (Costanza et al. 1997) as opposed to the total value of the goods and services themselves. The problem of having value estimates at the margin is that they should be used only to assess the changes in the condition of the environmental goods and services, and not their overall value (Rolfe and Windle 2009).

A number of sites in the Murray-Darling Basin have significant iconic values that people want to preserve whether or not they may use them. Some people are willing to pay to preserve such iconic sites for purely altruistic reasons, although Andreoni (1990) notes the existence of many other motivators. WTP can also be affected by the "warm glow" effect, whereby people derive a sense of moral satisfaction from the act of giving (Nunes and Schokkaert 2003; Kahneman and Knetsch 1992).

While care must be taken in conducting and interpreting nonmarket valuation studies, assembling the use and nonuse estimates is critical to understanding the magnitude of some resource management issues. Measuring the TEV of a river basin requires the knowledge of both use and nonuse values of the river. River-based recreational activities have the potential to generate significant amounts of income to the local community if the water quality, riverflow, and biodiversity values are improved or at least maintained. These values could be from both direct and indirect uses of the river. Many studies have investigated the use and nonuse recreational values in international settings.

Gibbs and Conner (1973) estimated the use and nonuse values of the Kissimmee River Basin in Florida. The use value derives from water-based recreational activities, which include fishing, water-skiing, boating, and swimming. Two types of recreationists were surveyed: those living on lakefront properties and those who travel to the area. The total use value from recreation was estimated to be US$29.2 million. The nonuse value is the recreational and aesthetic value of the river that is capitalized into riverfront residential properties. This value was estimated to be US$7 million.

Sanders et al. (1990) found the total value of protecting 15 rivers in Colorado to be US$120 million annually. Of this, only about one-fifth of the value was related to recreational use benefits, and the other four-fifths were for nonuse benefits. Douglas and Taylor (1998) evaluated the impacts of Trinity Dam in Northern California, including the loss of streamflow in the Trinity River and the adverse impacts on river-related recreational activities. The authors used the CVM and TCM to estimate the nonmarket benefits of improved streamflow in the Trinity River to Pacific Coast households. The CVM value reflects WTP to achieve a nearly pristine water and fishery resource, whereas the TCM value reflects the benefits for the current degraded state of the river and its aquatic resources. The WTP from the CVM technique suggests that households are willing to pay between US$17 million and US$42 million per year, while the TCM suggests that the current benefits are about US$406 million per year.

In the impaired Platte River Basin of Denver, Colorado, Loomis et al. (2000) measured the TEV of restoring ecosystem services. The benefits include the dilution of wastewater, erosion control, water purification, and improvement of wetland quality. Using the CVM, Loomis and colleagues estimated the total economic value to be between US$19 million and US$70 million.

More recently, in South Africa, Turpie and Joubert (2004) estimated the tourism value and potential changes in tourism value from changes in the water level of the Crocodile Catchment in Kruger National Park (KNP). The current tourism value was estimated using direct revenues generated by visitors' on-site expenditures, contribution to the economy from visitors' on-site and off-site expenditures, and recreational values from consumer surplus. A joint contingent and conjoint valuation approach estimated that the current value of KNP tourism from on-site expenditures is about A$20.5 million and A$40 million, respectively, in terms of economic contribution and A$150 million in terms of consumer surplus.[4] The authors approximated that about 30% of tourism business would be lost if rivers were totally degraded.

Examination of these studies reveals the substantial values associated with recreational and tourism uses of water and provides a backdrop against which to consider various water-sharing proposals. In the context of the Murray-Darling Basin, studies have also shown the often unrecognized value of these sectors.

SUMMARY OF VALUATION STUDIES RELEVANT TO THE MURRAY-DARLING BASIN

In an effort to identify the gaps in the state of knowledge about the value associated with the River Murray, Table 3.1 (overleaf) summarizes the various types of studies that have been completed on the recreation, tourism, and use or nonuse of natural assets in the Murray-Darling Basin. These studies include literature reviews, regional economic models, hedonic pricing models, travel cost models, and stated preference studies.

Literature reviews are instrumental in establishing a starting point for analysis of the issues. Hassall and Gillespie (2004) estimated the use value to be A$1.62 billion in a desktop review of the use value of healthy rivers in the Murray-Darling Basin.

Dyack et al. (2006) presented an economic analysis of the regional and sectoral implications of various environmental water scenarios in the River Murray Basin. Although some analysis has been done of the impact of changing water levels caused by environmental water diversions on tourism activity, there is little understanding of the extent to which improved water quality and environmental conditions might increase visitation. Dyack and colleagues focused on exploring the potential increases in regional value added that could follow from improved site conditions at tourism destinations along the River Murray from environmental flows of 125, 250, and 500 gigaliters. Using simple assumptions about demand and supply increases for most River Murray Basin regions, although a portion of the regional GDP is lost as a result of environmental water diversions, this could be offset in part by increases in tourism activity.

Table 3.1 Summary of studies inside and outside the Murray-Darling basin

Study	Technique	Sample	Location and asset	Values estimated in A$ (year of data collection)
Sinden (1990)	Travel cost survey	Recreationalists at 24 sites	Ovens and King Basin, Victoria	$22 for a day visit $37 for a camping visit
Carter (1992)	Travel cost survey	NSW, Victoria, and ACT households	Southeastern forests, NSW East Gippsland, Victoria	$8.90 per visitor $43.50 per person
Stone (1992)	Contingent valuation	Rural and urban areas in Victoria	Barmah, Victoria	$30 per person
Baker and Pierce (1998)	Contingent valuation	Adelaide River Murray	River Murray	$43 nonanglers—mean willingness to double native fish stocks $159 anglers—mean willingness to double native fish stocks
Bennett et al. (1998)	Contingent valuation	NSW households SA households	Tilley Swamp, SA Tilley Swamp and Coorong	$130 per household to avoid damage $200 per household to avoid damage
Morrison et al. (1999)	Choice modeling	Sydney households	Macquarie Marshes, NSW	$0.13–$1.14 irrigation-related employment $0.04–$0.05 wetland area $21.82–$24.62 frequency of waterbird breeding $4.04–$4.16 number of endangered species
Bennett and Whitten (2000)	Contingent valuation	Local residents	Lake Gol Gol and Gol Gol Swamp, NSW	$9.93 per household to prevent damage from rising salinity $20 per person who intends to visit
Whitten and Bennett (2001)	Choice modeling	Canberra Wagga Wagga Griffith Adelaide	Murrumbidgee River, NSW	$11.39 per 1,000 ha of wetland $0.55 per 1% native birds $0.34 per 1% increase in native fish populations $5.73 to prevent one more farmer leaving
Morrison (2002)	Choice modeling	Macquarie Valley households	Macquarie Marshes, NSW	$3.63 per extra 100 km² $8.98 per 1 year frequency of bird breeding $5.23 per extra endangered waterbird protected

Study	Method	Location	
Morrison et al. (2002)	Choice modeling	Sydney, NSW	Macquarie Marshes	$0.107 to support irrigation-related employment $0.034 per ha of wetland area $24.15 to increase frequency of bird breeding by 1 year $4.27 per endangered and protected species
		Sydney, NSW	Gwydir Wetlands, NSW	$0.218 to support irrigation-related employment $0.039 per ha of wetland area $9.81 to increase frequency of bird breeding by 1 year $3.21 per endangered and protected species
		Moree, NSW	Gwydir Wetlands, NSW	$0 to support irrigation-related employment $0 per ha of wetland area $15.18 to increase frequency of bird breeding by 1 year $3.86 per endangered and protected species
Morrison and Bennett (2004)	Choice modeling		*IN-CATCHMENT samples*	
		5 in-catchment samples	Bega River	$2.33 per 1% of river covered with healthy vegetation $7.23 number of fish species $100.98 suitable for fishing $51.33 suitable for swimming $0.88 per species
			Clarence River	$2.07 per 1% of river covered with healthy vegetation $0 number of fish species $72.77 suitable for fishing $46.63 suitable for swimming $1.92 per species

(continued)

Table 3.1 – continued

Study	Technique	Sample	Location and asset	Values estimated in A$ (year of data collection)
			Georges River	$1.51 per 1% of river covered with healthy vegetation $0 number of fish species $73.88 suitable for fishing $45.26 suitable for swimming $0 per species
			Murrumbidgee River	$1.98 per 1% of river covered with healthy vegetation $3.51 number of fish species $59.98 suitable for fishing $29.93 suitable for swimming $0 per species
		2 out-of-catchment samples	Gwydir River	$2.15 per 1% of river covered with healthy vegetation $4.05 number of fish species $86.46 suitable for fishing $28.75 suitable for swimming $1.79 per species
			OUT-OF-CATCHMENT samples Murrumbidgee River	$1.98 per 1% of river covered with healthy vegetation $3.51 number of fish species $59.98 suitable for fishing $29.93 suitable for swimming $0 per species

Reference	Method	Site	Location	Benefit
			Gwydir River	$2.15 per 1% of river covered with healthy vegetation
				$4.05 number of fish species
				$86.46 suitable for fishing
				$28.75 suitable for swimming
				$1.79 per species
Bennett et al. (2008)	Choice modeling	Goulburn River	In-catchment	$4.39 to increase % of presettlement fish species
				$3.56 to increase % of vegetation along river's length
				$3.90 to increase number of native birds and animals
				$2.12 to increase % of river suitable for primary contact
			Rural out-of-catchment	$5.56 to increase % of presettlement fish species
				$4.65 to increase % of vegetation along river's length
				$3.04 to increase number of native birds and animals
				$0 to increase % of river suitable for primary contact
			Melbourne	$4.47 to increase % of presettlement fish species
				$5.53 to increase % of vegetation along river's length
				$3.35 to increase number of native birds and animals
				$1.64 to increase % of river suitable for primary contact
Crase and Gillespie (2008)	Travel cost	Boaters	Lake Hume, NSW	$1.3 million per year in recreational benefits from raising levels from 50% to near full
Hatton MacDonald et al. (in press)	Choice modeling	New South Wales	River Murray and Coorong, Murray Darling Basin	$13.64 for a 1-year increase in waterbird frequency
				$2.50 for a 1% increase in native fish population
				$2.88 for a 1% increase in healthy vegetation
				$146.48 to improve waterbird habitat in Coorong

(continued)

Table 3.1 – *continued*

Study	Technique	Sample	Location and asset	Values estimated in A$ (year of data collection)
		Australian Capital Territory		$15.99 for a 1-year increase in waterbird frequency $3.58 for a 1% increase in native fish population $4.42 for a 1% increase in healthy vegetation $198.15 to improve waterbird habitat in Coorong
		Victoria		$12.00 for a 1-year increase in waterbird frequency $2.28 for a 1% increase in native fish population $2.87 for a 1% increase in healthy vegetation $126.63 to improve waterbird habitat in Coorong
		South Australia		$15.96 for a 1-year increase in waterbird frequency $2.15 for a 1% increase in native fish population $3.88 for a 1% increase in healthy vegetation $169.18 to improve waterbird habitat in Coorong
		Rest of Australia		$18.64 for a 1-year increase in waterbird frequency $1.71 for a 1% increase in native fish population $3.31 for a 1% increase in healthy vegetation $187.09 to improve waterbird habitat in Coorong
Rolfe and Dyack (2010)	Travel cost modeling Contingent behavior	Recreationalists at site	Coorong, SA	$111 per adult per visit day $17.20 willingness to pay for 1% increase in access
Dyack et al. (2007)	Travel cost modeling	Recreationalists at site	Barmah Forest, Victoria	$149 per adult per visit day
Outside Murray Darling Basin				
Whitten and Bennett (2001)	Choice modeling	Adelaide, SA Naracoorte, SA Canberra, ACT	Upper Southeast, SA	$0 per 1,000 ha per household wetland area $0.92 per 1,000 ha per household remnant vegetation $4.81 per species $0 ducks hunted (average per 1,000)

Study	Method	Subjects	Location	Results
Whitten and Bennet (2005b)	Estimate of producer surplus		Upper Southeast, SA	Value of new or expanded wetland use: $128,000 farm stay $206,000 specialized tours $168,000 mini-resort $205,000 charged day visits $1,000 camping $18,000 self-catering
Whitten and Bennet (2005a)	Travel cost	Duck hunters	Upper Southeast, SA	$47.73 per visit
Rolfe and Windle (2006)	Choice modeling	Brisbane	Condamine River, QLD	$238,000 value of "wetlands and organised shoot" $2.75 1% of healthy vegetation in the floodplain $0.08 to increase 1 km of waterways in good health $1.833 for % increase in number of protected Aboriginal cultural sites $3.42 for an increase in % of unallocated water
Rolfe and Prayaga (2007)	Contingent valuation	Anglers at 3 Queensland dams	Bjelke-Petersen Dam, Boondooma Dam, and Fairbairn Dam, QLD	Willingness to pay for 20% improvement in fishing experience: $19.02 Bjelke-Petersen $43.03 Boondooma $36.45 Fairbairn Dam
	Travel cost			Zonal travel cost model of regular visitors: $59.65 per person per trip to Bjelke-Petersen Dam $348.22 per person per trip to Boondooma Dam $904.40 per person per trip to Fairbairn Dam

(continued)

Table 3.1

Study	Technique	Sample	Location and asset	Values estimated in A$ (year of data collection)
Zander et al. (2010)	Choice modeling	Pooled sample across north and Australian southern cities	Fitzroy River, Daly River, and Mitchell River, QLD	$54 medium area of floodplain in good condition $124 large area of floodplain in good condition $74 for 3-star-quality fishing $126 for 4-star-quality fishing $162 okay condition—waterholes important to Aboriginal people $238 good condition—waterholes important to Aboriginal people $96 low income from irrigated agriculture $35 medium income from irrigated agriculture
Morrison and Hatton MacDonald (2011)	Compensating budget	South Australia	Upper Southeast, SA	$1,098 per ha wetlands $1,228 per ha grassy woodlands $1,080 per ha scrubland
Hatton MacDonald and Morrison (2010)	Choice modeling	South Australia	Upper Southeast, SA	$1,529 per ha wetlands $1,129 per ha grassy woodlands $810 per ha scrubland

Notes: A$1 = US$0.9978 as of January 2011; ACT = Australian Capital Territory; NSW = New South Wales; QLD = Queensland; SA = South Australia; ha = hectare; km = kilometer.

In the analysis, Murray Statistical Division (SD) in New South Wales, and Ovens-Murray SD in Victoria, are subject to the smallest declines in regional income when water is diverted to the environment. In these areas, the greatest potential for offsetting losses in agricultural income occur if tourism expands. The opposite is true for Mallee SD and Goulburn SD, in Victoria, and Murray Lands SD, in South Australia, where losses are greatest from water diversions and recoveries. To extend this work, more information is required on the relationships among the regional economy, hydrology, and environmental outcomes. This should include intertemporal links among rainfall, usage, environmental flows, reservoir stocks, and the regional economy.

Recent travel cost studies inside and outside the Murray-Darling Basin demonstrate that recreational values can vary substantially and the simple transfer of values to other recreation areas can be complex. Much depends on the potential for substitute recreational experiences. Rolfe and Prayaga (2007) estimate the annual value of three freshwater dams in Queensland outside the basin at A$0.9 million for Bjelke-Petersen, A$3.2 million for Boondooma, and A$4.5 million for Fairbairn. The low value associated with Bjelke-Petersen may be due to substitutes such as Boondooma. Dyack et al. (2007) estimate the total value of recreation to be A$57 million per year at the Coorong and A$13 million for Barmah-Millewa. The difference in total value is largely related to the total visitation expected at each site each year, as the value per trip per adult is similar.

The largest component of use value, according to Hassall and Gillespie (2003), is the amenity value of property. Howard (2008) outlines the growth in amenity living associated with inland retirement migration and recreational values associated with boating. Analysis by Tapsuwan et al. (2010) suggests that more is going on with the increasing value of lifestyle properties than just natural amenities. The trend toward moving to the country, sometimes called a "tree change," is based on the idea that urban residents are moving to rural towns to seek a more relaxed, laid-back lifestyle. If this is the case, the demand for environmental amenities, such as recreation in inland marinas, natural parks, and wetlands in rural areas, is expected to increase. This inevitably will have an impact on prices of rural land nearby. Tapsuwan and colleagues employed a hedonic pricing method in the South Australian Murray-Darling Basin on all rural property sales from 2006 to 2008. The study specifically estimates the effects of natural amenities, such as nature conservation parks, rivers, lakes, and wetlands, on property prices. Variables such as proximity from the property to the edge of the nearest natural park and park area were important. However, the level of "recreational attractiveness" of a park is proving to be a potentially important predictor of value as well. Measures of attractiveness are based on the facilities and activities offered, including information centers, toilets, parking areas, picnic and barbecue facilities, camping, fishing, and canoeing. The analysis suggests that the traditional explanatory variables commonly used in the hedonic analysis, such as distance to, or area of, natural amenities, may not always be a good predictor of property price increases. Overall, this highlights the interdependence of tourism, recreation, and quality of the natural environment in one of the larger sets of values.

A research project in progress by the Commonwealth Scientific Industrial Research Organisation (CSIRO) in Australia has been examining the trade-offs among environmental benefits of water upstream and downstream on the River Murray. The findings indicate that respondents are willing to pay substantial amounts to improve the quality of the River Murray and Coorong. The values in Hatton MacDonald et al. (in review) are larger when compared in real terms with those in previous studies of rivers in Australia (see, e.g., Morrison et al. 1999; Morrison and Bennett 2004). Overall, this suggests that preferences for restoring the health of the riverine environment may have strengthened over time.

CONCLUSIONS

Tourism, recreation, amenity, and nonuse values are substantial and may be growing over time. Understanding of these values will contribute important information to the water policy debate, particularly if water managers are to optimize the socioeconomic and environmental outcomes. The values summarized in Table 3.1 reveal a "patchiness" in the methods used and the assets studied. For instance, stated preference studies of medium to large iconic assets dominate the collection. A limited number of revealed preference studies have been undertaken in the basin. Six of 24 nonmarket studies identified apply the TCM, and only one was a hedonic pricing study. Revealed preference is clearly one area that is underresearched.

As is noted elsewhere in this book, before these values can be incorporated in hydroeconomic modeling of the basin to optimize multiple uses of water, it is important to explore the nature of the trade-offs among consumptive and nonconsumptive benefits conferred as a result of regulation and a more natural flow regime. Trade-offs among the major environmental assets at the bottom of the Murray-Darling Basin and upstream assets, amenities, and recreational activities at the top of the system are also poorly understood.

Overall, knowledge about the spatial nature of values is incomplete. Households across Australia are likely to hold nonuse values for the River Murray, and these values are likely to differ with location. People living within close proximity of the basin are likely to have higher use values than people living farther away, but care must be taken with respect to potential overlap with option values. Despite the complexities in evaluating the value of recreation in a water-constrained economy, we reiterate Howard's (2008) assertion that it is time that institutions improved their capacity to appreciate, understand, and incorporate both use and nonuse values when making water resource management decisions.

NOTES

1. In contrast, evidence from forestry suggests that the inclusion of recreational values can alter the management of a resource (Pearce et al., 2003; van Kooten and Bulte, 1999). Baker and Pierce (1997) demonstrate that the net social value of river fish stocks exceeds the market value 50 to 1.

2. For example, Ward et al. (1996) demonstrate how recreation values have altered with changing water levels as a result of drought. Loomis and Richardson (2001) suggest that economic values are associated with the protection of natural environments, specifically the wilderness in the United States, which include recreational benefits, passive-use benefits, and off-site benefits (i.e., the gain in property values surrounding the wilderness area).

3. A$1 = US$0.9978 as of January 2011.

4. These figures were derived using the September 2010 exchange rate for the South African rand (ZAR) and Australian dollar, whereby 1 ZAR = A$0.15.

REFERENCES

ABS (Australian Bureau of Statistics). 2009. *Australian National Accounts: Tourism Satellite Accounts*. cat. 5249.0. Canberra: Australian Government Publishing Service.

Andreoni, J. 1990. Impure altruism and donations to public goods: A theory of warm-glow giving. *Economic Journal* 100: 464–477.

Baker, D., and B. Pierce. 1997. Does fisheries management reflect societal values? Contingent valuation evidence for the River Murray. *Fisheries Management and Ecology* 4: 1–15.

Bateman, I., R. Carson, B. Day, M. Hanemann, N. Hanley, T. Hett, M. Jones-Lee, G. Loomes, S. Mourato, E. Özdemiroglu, D. Pearce, R. Sugden, and J. Swanson. 2002. *Economic Valuation with Stated Preference Techniques: A Manual*. Northampton, MA: Edward Elgar.

Bennett, J., R. Dumsday, G. Howell, C. Lloyd, N. Sturgess, and L. Van Raalte. 2008. The economic value of improved environmental health in Victorian rivers. *Australasian Journal of Environmental Management* 15: 138–148.

Bennett, J., M. Morrison, and R. Blamey. 1998. Testing the validity of responses to contingent valuation questioning. *Australian Journal of Agricultural and Resource Economics* 42: 131–148.

Bennett, J., and S. Whitten. 2000. *The Economic Value of Conserving/Enhancing Gol Gol Lake and Gol Gol Swamp. A Consultancy Report*. Canberra: Gol Gol Community Reference Group.

Boyd, J. 2007. Nonmarket benefits of nature: What should be counted in green GDP? *Ecological Economics* 61: 716–723.

Carter, M. 1992. The use of the contingent valuation in the valuation of national estate forests in south-east Australia, in *Valuing Natural Areas: Applications and Problems of the Contingent Valuation Method*, edited by M. Lockwood and T. De Lacy. Albury: Charles Sturt University, Johnstone Centre of Parks, Recreation and Heritage, 17–29.

Costanza, R., R. D'arge, R. De Groot, S. Farber, M. Grasso, B. Hannon, K. Limburg, S. Naeem, R. O'Neill, J. Paruelo, R. Raskin, P. Sutton, and M. van den Belt. 1997. The value of the world's ecosystem services and natural capital. *Nature* 387: 253–260.

Crase, L., and R. Gillespie. 2008. The impact of water quality and water level on the recreation values of Lake Hume. *Australasian Journal of Environmental Management* 15: 21–29.

Douglas, A., and J. Taylor. 1998. Riverine based eco-tourism: Trinity River non-market benefits estimates. *International Journal of Sustainable Development and World Ecology* 5: 136–148.

Dyack, B., E. Qureshi, and G. Wittwer. 2006. *Regional impacts of environmental water flows: Case study of tourism in the Murray River Basin*, paper presented at AARES Conference, February 8-10, 2006, Sydney.

Dyack, B., J. Rolfe, J. Harvey, D. O'Connell, and N. Abel. 2007. *Valuing Recreation in the Murray*. Canberra: CSIRO Water for a Healthy Country Flagship Program.

Gibbs, K.C., and J.R. Conner. 1973. Components of outdoor recreational values: Kissimmee River Basin, Florida. *Southern Journal of Agricultural Economics* 5: 239–244.

Hanley, N., J.F. Shogren, and B. White. 2001. *Introduction to Environmental Economics*. New York: Oxford University Press.

Hassall and Gillespie (Hassall & Associates Pty. Ltd. and Gillespie Economics). 2003. *Tourism and Recreation Economic Impact Scoping Study: The Living Murray Initiative*, report prepared for the Murray-Darling Basin Commission. Sydney: Hassall & Associates Pty. Ltd.

Hatton MacDonald, D., and M. Morrison. 2010. Valuing habitat using habitat types. *Australasian Journal of Environmental Management* 17: 235–243.

Hatton MacDonald, D., M. Morrison, J. Rose, and K. Boyle. In review. Valuing a multi-state river: The case of the River Murray. *Australian Journal of Agricultural Resource Economics.*

Horridge, M., J. Madden, and G. Wittwer. 2005. Using a highly disaggregated multi-regional single-country model to analyse the impacts of the 2002–03 drought on Australia. *Journal of Policy Modelling* 27: 285–308.

Howard, J. 2008. The future of the Murray River: Amenity re-considered? *Geographical Research* 46: 291–302.

Kahneman, D., and J.L. Knetsch. 1992. Valuing public goods: The purchase of moral satisfaction. *Journal of Environmental Economics and Management* 22: 57–70.

Krutilla, J. 1967. Conservation reconsidered. *American Economic Review* 57: 777–786.

Loomis, J., P. Kent, L. Strang, K. Fausch, and A. Covich. 2000. Measuring the total economic value of restoring ecosystem services in an impaired river basin: Results from a contingent valuation survey. *Ecological Economics* 33: 103–117.

Loomis, J., and R. Richardson. 2001. Economic values of the U.S. wilderness system: Research evidence to date and questions for the future. *International Journal of Wilderness* 7: 31–34.

Morrison, M. 2002. Understanding local community preferences for wetland quality. *Ecological Management and Restoration* 3: 127–132.

Morrison, M., and J. Bennett. 2004. Valuing New South Wales rivers for use in benefit transfer. *Australian Journal of Agricultural and Resources Economics* 48: 591–611.

Morrison, M., J. Bennett, and R. Blamey. 1999. Valuing improved wetland quality using choice modelling. *Water Resources Research* 35: 2805–2814.

Morrison, M., J. Bennett, R. Blamey, and J. Louviere. 2002. Choice modelling and tests of benefit transfer. *American Journal of Agricultural Economics* 84: 161–170.

Morrison, M., and D. Hatton MacDonald. 2011. A comparison of compensating surplus and budget reallocation with opportunity costs specified. *Applied Economics*, doi:1466-4283, January 28.

Nunes, P.A.L.D. 2002. *The Contingent Valuation of Natural Parks: Assessing the Warm Glow Propensity Factor.* Northampton, MA: Edward Elgar.

Nunes, P.A.L.D., and E. Schokkaert. 2003. Identifying the warm glow effect in contingent valuation. *Journal of Environmental Economics and Management* 45: 231–245.

Pearce, D., and D. Moran. 1994. *The Economic Value of Biodiversity.* London: Earthscan Publications Limited.

Pearce, D., F. Putz, and J. Vanclay. 2003. Sustainable forestry in the tropics: panacea or folly? *Forest Ecology and Management* 172: 229–247.

Rolfe, J. 2010. Valuing reductions in water extractions from groundwater basins with benefit transfer: The Great Artesian Basin in Australia. *Water Resources Research* 46, W06301, doi:10.1029/2009WR008458.

Rolfe, J., and B. Dyack. 2010. Valuing recreation in the Coorong, Australia, with travel cost and contingent behaviour models. *Economic Record*, doi:10.1111/j.1475-4932.2010.00683x.

Rolfe, J., and P. Prayaga. 2007. Estimating values for recreational fishing at freshwater dams in Queensland. *Australian Journal of Agricultural and Resource Economics* 51: 157–174.

Rolfe, J., and J. Windle. 2006. Valuing Aboriginal cultural heritage across different population groups, in *Choice Modelling and the Transfer of Environmental Values*, edited by J. Rolfe and J. Bennett. Cheltenham, UK, 216–244.

———. 2009. *A Systematic Database for Benefit Transfer of NRM Values in Queensland*, accessed February 8, 2011, from http://content.cqu.edu.au/FCWViewer/view.do?page=2598.

Sanders, L.D., R.G. Walsh, and J.B. Loomis. 1990. Toward empirical estimation of the total value of protecting rivers. *Water Resources Research* 26: 1345–1357.

Sinden, J.A. 1990. *Valuation of Unpriced Benefits and Costs of River Management: A Review of the Literature and a Case Study of the Recreation Benefits in the Ovens and King Basin.* Melbourne: Department of Conservation and Environment Victoria.

Stone, A. 1992. Contingent valuation of the Barmah Wetlands, Victoria, in *Valuing Natural Areas: Applications and Problems of the Contingent Valuation Method*, edited by M. Lockwood and T. De Lacy. Albury: Charles Sturt University, Johnstone Centre of Parks, Recreation and Heritage, 47–70.

Tapsuwan, S., D. Hatton MacDonald, and D. King. 2010. Valuing natural amenities in the South Australia Murray Darling Basin: A site recreation index approach in hedonic property pricing, paper presented at the *Fourth World Congress of Environmental and Resource Economists*, June 28–July 2, 2010, Montreal.

Turpie, J., and A. Joubert. 2004. Estimating potential impacts of a change in river quality on the tourism value of Kruger National Park: An application of travel cost, contingent and conjoint valuation methods. *Water SA* 27: 387–398.

van Kooten, C., and E. Bulte. 1999. How much primary coastal temperate rain forest should society retain? Carbon uptake, recreation, and other values. *Canadian Journal of Forest Research* 29: 1879–1890.

Ward, F., B. Roach, and J. Henderson. 1996. The economic value of water in recreation: Evidence from the California drought. *Water Resources Research* 32: 1075–1081.

Whitten, S., and J. Bennett. 2001. Non-market values of wetlands: A choice modelling study of wetlands in the Upper South East of South Australia and the Murrumbidgee River floodplain in New South Wales. *Private and Social Values of Wetlands Research Report No. 8*. Canberra: The University of New South Wales.

———. 2005a. Non-market use values of wetland resources. *Managing Wetlands for Private and Social Good: Theory, Policy and Cases from Australia*. Cheltenham, UK: Edward Elgar Publishing.

———. 2005b. Private value of wetlands. *Managing Wetlands for Private and Social Good: Theory, Policy and Cases from Australia*. Cheltenham, UK: Edward Elgar Publishing.

Young, M., W. Proctor, and G. Wittwer. 2006. Without water: The economics of supplying water to 5 million more Australians. In *Water for a Healthy Country Flagship Report*. Adelaide: CSIRO.

Zander, K., S. Garnett, and A. Straton. 2010. Trade-offs between development, culture and conservation—willingness to pay for tropical river management among urban Australians. *Journal of Environmental Management* 91: 2519–2528.

CHAPTER 4

Access to Inland Waters for Tourism: Ecosystem Services and Trade-offs

Pierre Horwitz and May Carter

*A*s discussed in the introduction to this book, in managing Australia's water resources, emphasis has been placed on securing drinking water and water for primary industry. However, as Hone noted in Chapter 2, the extended drought conditions affecting many areas of Australia have caused substantial degradation of its river systems. In this context, social and cultural values associated with water are receiving more attention (Pigram 2006). This is partly due to the recognition that social issues will inevitably arise from trade-offs in water allocation and management, and that water planning needs to be integrated with other natural resource management objectives (Hussey and Dovers 2007). Establishing some sort of balance between water allocation for industry, protection of drinking water quality, and access for tourism and recreational activities is emerging as a significant issue in the governance and management of freshwater resources in many parts of Australia (DERM 2007). Alongside this development, the management of water allocation and access is generating new models of collaboration, particularly in rural and regional areas where economic development has typically relied on agriculture or other primary industry.

This chapter explores several critical issues relating to management of inland freshwater systems and visitation for tourism and recreation, examining links among tourism, ecosystem services, and human well-being, and relationships between industry setting and visitor experience. Within this analysis, the concept of trade-offs is central, and we propose a framework that considers both how and why trade-offs among ecological, social, and economic effects might occur.

The chapter begins by providing a brief contextual description of inland freshwater systems and their visitation for tourism and recreation, followed by a proposal that visitation be seen from the perspective of deriving a variety of ecosystem services. Next we examine the systemic nature of ecosystem services, with the introduction of the concept that these services can be, and are, traded off when decisionmaking occurs over resource use, an example being between visitation and other water users. These concepts are then applied specifically: with the aid of a table, we describe the variety of aquatic settings that attract visitors and, across that range, the sorts of experiences and aquatic ecosystem features that have appeal for visitors. The next section shows, again with the aid of a set of tables, how each desired aquatic ecosystem feature and facility will be supported by, be provided for, or emanate from a specified set of ecosystem services, and how these services can be compromised. When compromise occurs, trade-offs are made in order to either maintain or forgo the visitor appeal. We conclude with some comments on why these trade-offs need to be recognized and how they can be negotiated.

INLAND FRESHWATER SYSTEMS AND VISITATION FOR TOURISM AND RECREATION

As noted in Chapter 1, water is essential to the aesthetic and functional appeal of water-based tourism and recreation (see also Curtis 2003; Pigram 2006). Tourism facilities with views of rivers and lakes are often cited as preferred environmental settings and are extremely popular landscape and cultural tourism destinations (Wahab and Pigram 1997). Involvement in water-based outdoor recreation and tourism activities can provide personal satisfaction and enjoyment and contribute to physical and mental health.

Despite water-based recreational and tourism activities being a major nonconsumptive use for water, relatively little data exists about user behavior or usage trends in Australia (Pigram 2006). The nuances of tourist behavior are the subject of Chapter 12. Limited research has been done on the sociocultural or health impacts of visitor access to freshwater resources or on the overall effect of diminishing water levels on the tourism industry. In general, the complexity of relationships between tourism and recreation and sustainable use of natural resources has not been well examined (McDonald 2009).

One exception is research relating to potentially detrimental effects on water quality through allowing access to water catchments for tourism and recreational activities. In fact, whether access to forested areas surrounding drinking water sources for recreation and tourism activities should be allowed is being fiercely contested of late (Davison et al. 2008; Gray 2008; Hughes et al. 2008; Krogh et al. 2009). Apart from public health risks associated with microbiological contamination, access for tourism and recreational activities can also result in detrimental ecological effects. These effects can include vegetation trampling, soil erosion, harm to wildlife, introduction of pests and diseases, and fire (Krogh et al. 2009).

Human-made water features also provide opportunities for tourism and recreation. Many drinking water reservoirs constructed in the early 1900s were built with parks and gardens on their embankments to provide communities with recreation close to water. Over ensuing decades, water reservoirs built in Perth, Canberra, Melbourne, and Sydney were located within forested areas that served as natural buffers. As further protection, exclusionary policies prohibited public access to some water sources and surrounding forest (Krogh et al. 2009). Pressure on authorities has increased to open water catchments to meet demand for access to forested areas and settings that provide opportunities for water-dependent and water-enhanced tourism and recreational activities (Pigram 2006). Currently there is little consistency across Australia with regard to tourism-related or recreational access or exclusion from freshwater sources and their catchments (Hughes et al. 2008; Leisure Futures 2009); these facets are explored in detail in Chapter 9. In this context, examining the way the services provided by ecosystems are contested by tourism and other water users holds some promise.

The Millennium Ecosystem Assessment (MA) demonstrated that changes in ecosystem services influence human health and proposed the actions needed "to enhance the conservation and sustainable use of ecosystems and their contributions to human well-being" (2005, ii). Ecosystem services provide benefits and resources to people. These might include supporting services such as soil formation and nutrient cycling; provisioning services such as food and fresh water; regulating services such as climate regulation and water purification; and cultural services such as recreational, aesthetic, and spiritual benefits (Figure 4.1).

Six types of cultural services provided by ecosystems are identified within the MA: cultural diversity and identity; cultural landscapes and heritage values; spiritual services; inspiration (such as for arts and folklore); and recreation and tourism (MA 2005). Two of these services, exploring cultural diversity and

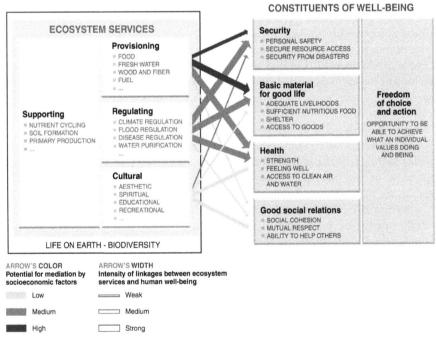

Figure 4.1 *Linkages between ecosystem services and human well-being*

Source: MA (2005)

landscape and heritage, have strong links with tourism, and the complex links between human health and ecosystem health need to be considered (Chivian 2002; Neller 2000; Verrinder 2007; Wilcox 2001). Many health-related benefits can be derived from the use of cultural ecosystem services for aesthetic appreciation, spiritual connection, and participation in educational and recreational activities. These include enjoying nature and escaping civilization; escape from routine and responsibility; creativity and self-improvement; relaxation; social contact and meeting new people; stimulus seeking; self-actualization (self-improvement and ability utilization); and challenge, achievement, and competition (Ibrahim and Cordes 2008).

Access to fresh water is identified as a key ecosystem service, and use of this resource is now considered to be well beyond sustainable levels, currently and into the future (MA 2005). Tourism is dependent on the availability of good-quality water, and water-rich areas have high tourist appeal (Pigram 2006). High tourism demand can place corresponding levels of demand on water resources and contribute to their degradation. Problems of water scarcity can be exacerbated by increased demand, not only for drinking water, but also for water for swimming pools, landscaping, and golf courses; visitor expectations of higher water quality; conflict with traditional land uses and communities over scarce supply; and limited options for disposal and treatment of water containing wastes, with potential for pollution or groundwater contamination. A full discussion of these issues is the subject of Chapter 13.

Competition between and within consumptive and nonconsumptive water uses suggests that some ecosystem services are being advanced at the expense of others. Even within the tourism industry and between different forms of visitation, trade-offs are constantly being made.

TRADE-OFFS, ECOSYSTEM SERVICES, AND THE TOURISM INDUSTRY

The concept of water trade-off is becoming more established within areas experiencing water scarcity. Trade-offs can be broadly defined as occurring in situations where the human demand on a natural resource exceeds the supply of that resource, so that there needs to be negotiation among the resource users to determine how that resource is to be shared. To achieve often unstated desirable outcomes, various demands on water sources are traded off, one at the expense of another.

Common trade-offs occur between water allocation and use for consumption (urban or irrigation) and environmental flows or water for tourism and recreational purposes; and between land uses that provide a range of services to humans (such as agriculture, forestry, urban development) and maintaining biodiversity. Competition also occurs in the demand for water-based or land-based resources at the expense of one over another (Molden 2007). A key category of trade-off of particular interest for this chapter is where ecosystem services (for particular health or well-being gains) are disrupted because another set of ecosystem services is enhanced to produce different health or well-being gains. A conspicuous example is where a provisioning service (using water from a wetland to benefit human well-being, such as extraction of water from an aquifer for agriculture, yielding livelihood benefits and food production) is enhanced at the expense of regulating services (those that might have negative consequences for human health, such as where acid sulfate soils are exposed because of lowering of water tables, mobilizing heavy metals and resulting in their incorporation into food chains and direct human exposure) (Horwitz et al. 2011).

Rodriguez et al. (2006) classify such ecosystem service trade-offs along three axes: a spatial scale, concerning the degree to which the effects of the trade-off are felt locally or at a distant location; a temporal scale, referring to whether the effects take place relatively rapidly or slowly; and the degree of reversibility, which expresses the likelihood that an ecosystem service might be extinguished and unable to return to its original state if management regimes prioritize other ecosystem services. Trade-offs are invariably part of a social ecological system where complex patterns of events result in either virtuous circles (where societal benefits reinforce decisionmaking, which continues to generate societal benefits) or vicious circles (where decisionmaking makes conditions in the system worse, which reinforces decisionmaking). A virtuous circle can transform into a vicious circle if eventual negative feedback is ignored or traded off. The "pathology of natural resource management" first described by Holling and Meffe (1996) is a good example: natural resource management policies and development that initially succeed become focused on efficiency, leading to agencies that become entrenched in their ways. In doing so, they gradually become more rigid and myopic, ignoring or trading off signals showing that economic sectors have become dependent on them, that ecosystem services are being depleted and ecosystems are becoming more fragile, and that the public is losing trust in governance. A collapse of the resource or another form of crisis eventuates. The virtuous circle switches to a vicious circle, hinged on the ability to detect and respond to the signals. These abilities are not developed through simple formulas; rather, they require flexible and adaptive approaches of awareness and response.

Trading off ecosystem services is neither uncommon nor unproblematic, and decisionmakers rarely see the full picture of ecosystem services, nor do they weight them appropriately. To compound these matters, ecosystem services are systemically linked, they may not be independent of each other, and few relationships between services are linear (Rodriguez et al. 2006).

An example of how trade-offs might occur between ecosystem services and tourism development is provided by Essex et al. (2004), where management of ongoing problems with water supply on the Spanish island resort of Mallorca meant that ecological, social, and economic trade-offs were implemented in order to maintain visitation levels. This island attracts some 5.5 million visitors during the high season and is typical of many Mediterranean mass tourism destinations with warm climate and low rainfall. These environmental conditions contribute to water supply issues, with problems further compounded by above-average consumption of water by tourists compared with local residents. During the late 1990s, water was imported from Spain, though this was aborted because of the high cost and opposition from the people living in the Ebro Valley, where the water was being sourced. Ebro Valley residents objected to extraction of "their" water supplies and subsequent environmental damage to regional wetlands. In 2001, an "ecotax" to fund environmental conservation and restoration was imposed on visitors. Establishment of a desalination plant and changes in water management policy, particularly use of wastewater and recycled water for irrigation of golf courses, along with water metering, water audits, and public awareness and education campaigns, have lessened some water-related problems for Mallorca's tourism industry. There is still concern, however, that continuing overextraction from aquifers is resulting in increasing saline intrusion and saltwater contamination of surface-water sources (Essex et al. 2004). An interpretation of this is that some ecosystem services are being traded off, derived at the expense of others, to support fixed visitation levels, entrenching a vicious circle.

To avoid similar scenarios, it is imperative that those involved in the development of visitor facilities and services consider water availability in the early stages of planning (Pigram 2006); implement strategies that address economic, social, and ecological sustainability (Essex et al. 2004); and design monitoring processes capable of detecting signals from the system (Holling and Meffe 1996). Examining the systemic nature of ecosystem services, as well as trade-offs among ecosystem services, provides a vehicle for doing so. An important first step in this regard is to consider the variety of aquatic settings that attract visitors and, across that range, the sorts of experiences that visitors seek from aquatic ecosystems. Both will allow for the explicit recognition of ecosystem services required for visitation.

SETTINGS AND VISITOR EXPERIENCES

Visitation to freshwater systems can affect water quality in various ways. It is important to protect water resources and prevent ecological damage through environmental exploitation. However, although establishing protected areas or restricting human interaction may assist in maintaining some aspects of ecosystem health, exclusionary policies may have unintentionally detrimental effects on local populations through loss of access to local food supplies or socially or culturally important sites (MA 2005).

Interventions such as controlling access or hardening water edges to reduce erosion and mitigate ecosystem degradation may also reduce or improve the attractiveness of particular tourism destinations. For some tourists, evidence of ecological damage and human intervention through built environment changes may substantially detract from the experience they seek. For others, installation of visitor services and amenities may well contribute to heightened experience through ease of access or perceptions of lower risk (Pigram 2006).

To better manage sites where visitation occurs, some water managers have begun to use classification systems that assess landscape appearance, from undeveloped to developed, and the presence or lack of facilities (DERM 2007). Classification of settings is designed to assess the diversity and range of recreational opportunities within a regional area. It is recognized that visitors may have

Table 4.1 *Visitation for tourism and recreation: Industry settings for aquatic ecosystem services*

Degree of development	Undeveloped 1	2	3	4	5	6	7	8	Developed 9
Industry setting	Wild and remote places	Limited access, few trails	Rough trails, remote camping	Vehicle access tracks are main facility	Formed tracks with cleared areas	Developed campgrounds	Parks, picnic shelters	Highly modified with visitor amenities	Urban, industrial, no recreation amenities
Visitor activities	Wilderness hiking and camping, swimming, rafting and paddling, communing with nature			Bushwalking, 4WD, fishing, paddling, some small craft boating and fishing, some overnight camping		Overnight and extended camping, boating and fishing, walking, relaxing, mountain biking, picnicking, sight-seeing		Recreation and/or cultural attractions	
Visitor facilities and social expectation	Limited access, sense of isolation, expect few facilities or amenities			Vehicle access, expect to see other visitors, limited facilities and amenities		Easy access, expect to see many other visitors, infrastructure to support recreational activity		Easy access, many visitors, well-developed facilities	
Aquatic ecosystem features that have appeal for visitors (see Table 4.3)	*High appeal* Water quantity (potable water), clarity (clear), temperature (seasonal relevance), views, high (fresh) water quality. No toxicants and pathogens (or signs of human defecation) Riparian vegetation, locally characteristic plants and animals *Desirable but not essential* Recreational fishing (native species)			*High appeal* Water quantity (potable water), clarity (clear), temperature (seasonal relevance), views, quality (no odors) No toxicants and pathogens Emergent plants in water, overhanging vegetation, riparian shade, locally characteristic plants and animals Recreational fishing (native or introduced species) *Desirable but not essential* Accessibility to water bodies (short and safe when desired)		*High appeal* Water temperature (seasonal relevance), views, quality (no odors) No toxicants and pathogens Accessibility to water bodies (short and safe when desired) Land-based infrastructure (water, toilets, shelters) Other visitors Recreational fishing (native or introduced species) *Desirable but not essential* Water clarity (clear) and quantity (potable water) Emergent plants in water; overhanging vegetation, riparian shade, locally characteristic plants and animals		*High appeal* Water clarity (clear), views, quality (no odors) No toxicants and pathogens Absence of tree stumps in water bodies Accessibility to water bodies (short and safe when desired) Land-based infrastructure (water, toilets, shelters) Other visitors	

Source: Adapted from DERM (2007)

different expectations and seek different experiences within specific settings. Indeed, different settings attract different types of visitors.

Table 4.1 provides descriptions of various industry settings, from undeveloped to developed; the types of activities that might occur in each setting; and the expectations that visitors might have in relation to built facilities and social interaction. The aquatic ecosystem features that contribute to visitor appeal in each setting are also described.

Visitor experience of water-based ecosystems, particularly for those tourists seeking nature-based or ecotourism experience, is influenced by the quality of sensory and emotive responses, with the sights, smells, sounds, and feel of the landscape being visited all playing a part (Pigram 2006). Changes to water quantity, clarity, or temperature may result in lessened visual aesthetics through decreased water levels or discolored or cloudy water. Degradation of riparian vegetation may also affect shoreline appearance, and erosion may make water bodies more difficult to access. Higher water temperatures may make swimming more comfortable but can have devastating effects on water quality through proliferation of weeds or algae and associated odors. The presence of wildlife is another important factor that can positively or negatively influence visitor experience, depending on whether resident wildlife is seen as attractive (birds, small mammals) or problematic (mosquitoes). At another level, changes to water quantity and quality may impede participation in recreational activity, particularly water sports such as boating, canoeing, water-skiing, or swimming (Hadwen et al. 2008a, 2008b). This leads to two important points: each experience or aquatic feature draws on one or more ecosystem services as benefits, and each experience or aquatic feature can degrade or deplete one or more of those services. Table 4.2 provides a list of aquatic ecosystem services and whether they are cultural, provisioning, regulating, or supporting; relevant for visitation; and associated with water features (Table 4.3a), biodiversity (Table 4.3b), and visitor facilities (Table 4.3c) found in wetland ecosystems.

VISITOR EXPERIENCES, FEATURES OF AQUATIC ECOSYSTEMS, AND ECOSYSTEM SERVICES

In many cases, it is possible to map the types of ecological disturbances associated with visitation (Hadwen et al. 2008a, 2008b; Krogh et al. 2009). Disturbance of soils and sediments occurs where people walk, ride, or drive and often results in track widening, deepening, and erosion. Camping can lead to site compaction and destruction of vegetation. Motor vehicles and powerboats can contaminate land and water systems through fuel spills, oil and grease discharges, and engine operation. Fishing can exacerbate decline of freshwater species, and fish stocking of lakes and streams can result in domination of introduced species. Litter and waste pollution are likely to occur with visitation for any purpose.

Common management strategies include controlling the number of visitors to particular sites; providing designated paths or diverting traffic away from fragile areas; building shoreline facilities such as jetties, lookouts, boardwalks, or other structures; restricting vehicle or boating access; and providing waste cans and

Table 4.2 *List of aquatic ecosystem services*

Ecosystem services	Water features (4.3a)	Biodiversity (4.3b)	Visitor facilities (4.3c)
Cultural			
Science and education values		●	
Cultural heritage and identity		●	
Contemporary cultural significance		●	●
Aesthetic and sense of place values	●	●	●
Spiritual, inspirational, and religious values	●	●	●
Water-based sport and recreation	●		
Provisioning			
Water for drinking (humans or livestock)	●		
Water for industry	●		
Food for humans		●	
Other products and resources, including genetic material		●	
Regulating			
Groundwater replenishment	●	●	
Water purification/waste treatment or dilution	●	●	
Biological control agents for pests/disease		●	
Flood control, flood storage	●	●	
Coastal shoreline and riverbank stabilization and storm protection	●	●	●
Other hydrological services	●	●	
Local climate regulations/buffering of change	●	●	
Carbon storage and sequestration		●	
Hydrological maintenance of biogeochemical processes	●	●	
Supporting			
Nutrient cycling	●	●	
Primary productivity		●	
Sediment trapping, stabilization, and soil formation	●	●	

Note: This is not a complete list of ecosystem services; see Ramsar Convention (2008)

toilets (Hadwen et al. 2008a; Hughes et al. 2008). However, some indirect and associated impacts on ecosystem services are not as easily managed. Increased turbidity, eutrophication, toxic exposures, weed infestation, or presence of exotic species can result from human visitation and disturbance of local ecosystems. Although not all of these impacts may directly affect water quality, they highlight ecological vulnerabilities and potential degradation of catchment systems (Krogh et al. 2009). Systemic (negative) feedbacks occur as well; for instance, visitors seek to benefit from cultural ecosystem services, but overuse in popular areas produces concomitant impacts that result in the loss of supporting, provisioning, or regulating ecosystem services, and the cultural ecosystem services associated with tourism may in time become substantially reduced.

Table 4.3 examines selected relationships between changes to ecosystem services that may influence visitor experience. Emphasis is placed on exploring relationships between visitor experience and changes to water features (Table 4.3a), among biodiversity features (Table 4.3b), or through development of visitor facilities (Table 4.3c). Each table shows the processes that compromise the

Table 4.3a *Water features associated with wetland ecosystems likely to be relevant for visitation*

Aquatic ecosystem features	Processes that change features (compromise the service)	Relationships between visitor experience and changes in features	Trade-offs made among ecosystem services
Water			
Absence of toxicants	Increased dissolved organic carbon, sediment, and nutrients from catchment disturbance	Potential for state change from clear macrophyte dominated to turbid plankton dominated. If accompanied by nutrients or shoreline decay, there may be elevated perceptions of risk related to perceptions of ill health.	Choices made to enhance cultural services where overuse or inappropriate use comes at the expense of supporting and regulating services. This might occur as a redistribution of services: trading access (cultural) for water quality (regulating)
Clarity (clear water)	Discharge that overrides assimilatory and regulatory capacities; change to biogeochemical processes that results in mobilization of toxicants		
Quality (absence of odors)	Drought with loss of volume and exposure of previously anaerobic sediments	Decrease in aesthetic value and appeal may lead to decrease in visitation/recreation opportunities, particularly a reduced number of swimmers	A consequence of these types of trade-offs might be that they eventually undermine cultural services themselves, and that tourists go elsewhere (trading one tourist destination for another, where the process might be repeated)
Quantity (availability of water—consumptive)	Eutrophication (nutrient delivery and oversupply) Not enough water from drought and over-extraction	Insufficient water may make tourist facility unviable, with possible consequence of moving infrastructure to where water view continues	If water is overextracted, provisioning services are enhanced at the expense of cultural services
Temperature (seasonal/aseasonal relevance)	Removal of riparian vegetation and shade Sediment delivery—erosion from access points and trails	Removal of riparian vegetation and shade may increase water temperatures and enhance appeal for water-based activities	Switching can occur among regulating services, where some are enhanced at the expense of others, and this can result in a change in ecosystem character, where particular ecosystem states are "chosen" over others
Views	Shoreline deposition of decaying algae/macrophytes Too much water from upstream (where flood control/storage compromised)		

Table 4.3b Biodiversity features associated with water and wetland ecosystems likely to be relevant for visitation

Aquatic ecosystem features	Processes that change features (compromise the service)	Relationships between visitor experience and changes in features	Trade-offs made among ecosystem services
Biodiversity			
Absence of vector-borne and water-borne pathogens	Changes in riparian zone structure and function, and delivery of large woody debris (LWD)	Bird-watchers and recreational fishers may respond negatively to any loss of aquatic vegetation (habitat) or loss of LWD and associated habitat	Trade-offs are multidirectional and context-specific and occur among different consumptive and nonconsumptive uses of biodiversity features
Emergent plants in water, overhanging vegetation, riparian shade	Desnagging activities by visitors or resource managers	Visitors may respond positively to desnagging activities or removal of aquatic plants, particularly if they want to partake in water-based activities	Where aquatic habitats and their species are manipulated to enhance cultural services, some regulating services (e.g., flood storage; soil, sediment, and nutrient retention) are enhanced at the expense of others (e.g., water purification/treatment; species trophic interactions)
Presence or absence of tree stumps in water	Changing aquatic habitat to suit particular forms of visitation	Enhanced visitation where fish stocks are present or sustained or vice versa; visitation likely to be event-related or seasonal; fishing bans in some areas may significantly reduce visitor loads	Enhancing cultural services associated with fishing means eroding (trading off) regulating services (particularly species trophic interactions, biological control agents) and perhaps even other cultural services (like educational or nature study)
Locally characteristic plants and animals	Increased nutrient loads and changed flow regimes, which can influence aquatic macrophyte growth and abundance	Changes in water quality likely to lead to reduced visitation; certainly expect a reduced number of swimmers	If species are introduced, there is a trade-off between the ability of native species to provide these services and those abilities of introduced species; overtaking of fish stocks means that cultural services are exchanged for several supporting and regulating services
Recreational fishing (native species)	Disturbances associated with overuse (too many visitors), which can reduce species diversity and adversely influence animal behavior or result in introduction of weeds or other non-native species	Visitors are attracted to protected areas to experience native character of an area; human-wildlife interactions can increase wildlife abundance (of some species) but decrease diversity; relationships may be strongly seasonal	
Introduced fish	Increased fishing opportunities in some areas in response to fish stocking		
	Changes to fish population dynamics through overfishing; altered reproductive productivity; species interactions (predation, competition, diseases); changed aspects of dispersal		
	Emergent phenomena possible as a result of new biological relationships, including new or chance human exposures		

Table 4.3c *Visitor facilities associated with water and wetland ecosystems likely to be relevant for visitation*

Aquatic ecosystem features	Processes that change features (compromise the service)	Relationships between visitor experience and changes in features	Trade-offs made among ecosystem services
Visitor facilities			
Accessibility to water bodies (short and safe when desired)	Erosion of landform features	Too few, just right, or too much: provision of features and facilities, or the presence of other people, will be perceived positively until an individually determined point is reached, after which perceptions will be negative	Increased visitor numbers can lead to a need to change accessibility (spatial and temporal) at key sites; supporting and regulating services are exchanged for cultural services; trade-offs among access, demand, and increases in visitor numbers may occur at key sites
	Access facilitated through geomorphological modifications (rock, sediment, water; infrastructure like jetties, boardwalks), with disturbances changing water-sediment interactions and providing for intrusions of other organisms	Some visitors will respond positively to improved access, for boating in particular	
Other visitors, site personnel		Site "hardening" can reduce the potential impacts of visitor use on focal sites; some visitors do not like hardening, as it detracts from wilderness experience, but even so, provision of boardwalks for bird-watching and other uses is a popular application of sustainable management	
Land-based infrastructure	Access point modifications and changing facilities influencing the use and visitor loads at key sites	Some visitors to protected areas prefer relatively inaccessible sites, so changes in accessibility may detract from overall appeal; the higher the degree of disturbance (litter, water quality, etc.), the more likely visitors are to select alternate sites	"Improved" access is traded off against increased numbers of tourists (mainly swapping cultural services: improved access to educational services might degrade spiritual or aesthetic values)
	Antisocial behaviors created by the nature of surroundings and facilities, the clientele encouraged to the site, or lack of respect for other aquatic features	Spatial use and impacts of key sites are likely to spread in response to increased visitor loads, as some visitors try to get away from the crowds; social encounters can be part of satisfying encounters; numbers attract to a certain point and the presence of site personnel can be positive for security reasons up to a certain point; for some visitors, numbers of others make no difference	
	Too few or too many site personnel detracting from visitation experience	Popular sites may provide more facilities (toilets, showers, boardwalks, parking lots); when visitors are involved in water-based activities, the provision of land-based facilities can be important	

Source: Parts of the table (processes that change the feature and visitor relationships) have been adapted from Hadwen et al., (2008a, 2008b); ecosystem services that support, provide, or emanate from the features are shown in Table 4.2)

ecosystems, the ways visitors might react to that compromise, and, where the compromise is by choice, the trade-offs that need to be made explicit. Each feature and facility will be supported by, provided for, or emanate from a specified set of ecosystem services. These services can be compromised, in which case trade-offs are made in order to either maintain or forgo the visitor appeal. Many of the trade-offs or compromises described are accepted as part of natural resource management practice. Apart from setting specific trade-offs that may be instigated by site managers, tourists may initiate their own form of trade-off by avoiding one site because of lessened appeal and choosing to visit another that better meets their expectations of desired experience. The potential cycle of use, overuse, ecosystem decline, intervention, and transference of visitors to new sites is an ongoing issue within tourism management and for those charged with management of natural resource settings where visitation occurs.

CONCLUSIONS

In this chapter, we have provided an analytical framework for the tourism sector to input its requirements into policy development for water resource management. It builds on a categorization of visitation settings, the features of an ecosystem that appeal to visitors in each setting, and the ecosystem services that are subsequently derived in each case. From the perspective of the tourism sector in its broadest sense, and as outlined Table 4.1, Tables 4.2 and 4.3 therefore illustrate five properties of aquatic ecosystems that have remained unconnected until now:

- aquatic ecosystem features that are acknowledged as having appeal for visitors;
- ecosystem services that match these desirable features and are therefore relevant for visitation;
- processes that change the feature (by compromising the ecosystem service), such as environmental or social impacts;
- relationships among visitor experience, visitor behaviors, and changes in aquatic ecosystem features; and
- the trade-offs made when ecosystem services relevant for visitation are compromised.

These properties can be explored in any local or regional setting where visitation occurs, at any stage of the planning or policy process. The multiple perspectives gained from a full spectrum of the stakeholders involved in visitation will enrich the outcome.

The challenge will then be to convert these relatively abstract trade-offs into management responses. This could be achieved via several quite different mechanisms, most of which are beyond the scope of this chapter, but essentially the value of this analysis will center on developing an understanding of the systemic nature of these trade-offs, describing decisions by acknowledging trade-offs, and developing both nonmarket- and market-based approaches that take them into account.

Where there are trade-offs, it is important for politicians, regulators, and the public to understand the consequences of taking one path in preference to another (Rodriguez et al. 2006). Recognizing the potential for trade-offs is the important first step in this understanding, and modeling consequences under different scenarios, for each of the scales outlined by Rodriguez and colleagues, will be part of that process. Like others (see Campbell et al. 2010), we acknowledge that recognizing trade-offs is no simple matter, because a trade-off perceived by one stakeholder may not be recognized by another. So undertaking a process by which the trade-offs and their consequences are fairly negotiated becomes the central concern: representation of marginalized stakeholders, increased transparency of information, and engaging with the core pursuits of other sectors will be key components of such a process. As Dahlberg and Orlando (2009) say, "For a trade-off to be accepted in the long term, it has to be transparent and regarded as the outcome of fair negotiations. This does not mean that everyone concerned is satisfied with all aspects of an agreement, but that the process is regarded as just by the majority." The theoretical and conceptual aspects of this type of approach are investigated in detail in Chapter 11.

Ecosystem services can provide a common language for such dialogue; where excessive use of ecosystem services occurs, making trade-offs explicit will allow for negotiations regarding compensation or elimination of subsidies (and perverse incentives). Transfer of these subsidies to payments for nonmarketed ecosystem services also becomes possible. Payments for environmental services in the management of wetlands can address difficult trade-offs by bridging the interests of different stakeholders through compensations (see Wunder 2007), where again the question of how such negotiations take place becomes critical. Some of these issues are further addressed in Chapter 6.

REFERENCES

Campbell, B.M., J.A. Sayer, and B. Walker. 2010. Navigating trade-offs: working for conservation and development outcomes. *Ecology and Society* 15 (2): 16.

Chivian, E. (Ed.), 2002. *Biodiversity: Its importance to human health (Interim Executive Summary)*. Boston: Harvard Medical School.

Curtis, J.A. 2003. Demand for water-based leisure activity. *Journal of Environmental Planning and Management* 46 (1): 65–77.

Dahlberg, A.C., and C. Burlando. 2009. Addressing trade-offs: Experiences from conservation and development initiatives in the Mkuze wetlands, South Africa. *Ecology and Society* 14 (2): 37.

Davison, A., D. Deere, and P. Mosse. 2008. *Practical guide to understanding and managing surface water catchments*. Shepparton, VIC: Water Industry Operators Association of Australia.

DERM (Queensland Department of Environment and Resource Management). 2007. *Landscape Classification System for Visitor Management*, accessed September 7, 2010, from www.derm.qld.gov.au/services_resources/item_list.php?series_id=205386.

Essex, S., M. Kent, and R. Newnham. 2004. Tourism development in Mallorca: Is water supply a constraint? *Journal of Sustainable Tourism* 12 (1): 4–26.

Gray, N.F. 2008. *Drinking Water Quality: Problems and Solutions*. Cambridge, UK: Cambridge University Press.

Hadwen, W.L., A.H. Arthington, and P.I. Boonington. 2008a. *Detecting Visitor Impacts in and Around Aquatic Ecosystems within Protected Areas*, Sustainable Tourism Cooperative Research Centre: Griffith University.

Hadwen, W.L., W. Hill, and C.M. Pickering. 2008b. Linking visitor impact research to visitor impact monitoring in protected areas. *Journal of Ecotourism* 7 (1): 87–93.

Holling, C.S., and G.K. Meffe. 1996. Command and control and the pathology of natural resource management. *Conservation Biology* 10: 328–337.

Horwitz, P., M. Finlayson, and P. Weinstein, 2011. *Healthy Wetlands, Healthy People: A Review of Wetlands and Human Health Interactions*: Ramsar Technical Report, Ramsar Convention on Wetlands, Gland, Switzerland.

Hughes, M., M. Zulfa, and J. Carlsen. 2008. *A Review of Recreation in Public Drinking Water Catchment Areas in the Southwest Region of Western Australia*. Perth: Curtin Sustainable Tourism Centre, Curtin University.

Hussey, K., and S. Dovers. 2007. *Managing Water for Australia: The Social and Institutional Challenges*. Collingwood, VIC: CSIRO Publishing.

Ibrahim, H., and K.A. Cordes. 2008. *Outdoor Recreation: Enrichment for a Lifetime*. Champaign, IL: Sagamore.

Krogh, M., A. Davison, R. Miller, N. O'Connor, C. Ferguson, V. McClaughlin, and D. Deere. 2009. *Effects of Recreation Activities on Source Water Protection Areas: Literature Review.* Melbourne: Water Services Association of Australia.

Leisure Futures. 2009. *Case Examples of Recreation in and Around Australian Public Drinking Water Sources and Their Catchments*. Brisbane: Seqwater.

MA (Millennium Ecosystem Assessment). 2005. *Ecosystems and Human Well-being: Synthesis*. Washington, DC: Island Press.

McDonald, J.R. 2009. Complexity science: An alternative world view for understanding sustainable tourism development. *Journal of Sustainable Tourism* 17 (4): 455–471.

Molden, D. (Ed.), 2007. *Water for Food, Water for Life: A Comprehensive Assessment of Water Management in Agriculture*. London, Earthscan.

Neller, A.H. 2000. Opportunities for bridging the gap in environmental and public health management in Australia. *Ecosystem Health* 6 (2): 85–91.

Pigram, J.J. 2006. *Australia's Water Resources: From Use to Management*. Collingwood, VIC: CSIRO Publishing.

Ramsar Convention. 2008. *Ramsar Convention.* Resolution X.15: Describing the ecological character of wetlands, and data needs and formats for core inventory: Harmonized scientific and technical guidance. Ramsar Convention on Wetlands, Gland, Switzerland.

Rodriguez, J.P., D.T. Beard Jr, E.M. Bennett, G.S. Cumming, S.J. Cork, J. Agard, A.P. Dobson, and G.D. Peterson. 2006. Trade-offs across space, time and ecosystem services. *Ecology and Society* 11 (1): 28.

Verrinder, G., 2007. "Engaging the health sector in ecosystem viability and human health: What are barriers to, and enablers of, change?," in *Ecology and Health: People and Places in a Changing World*, Edited by P. Horwitz, Melbourne, Organising Committee for the Asia-Pacific EcoHealth Conference 2007, 25–29.

Wahab, S., and J.J. Pigram. 1997. *Tourism, Development and Growth: The Challenge of Sustainability.* New York: Routledge.

Wilcox, B.A. 2001. Ecosystem health in practice: Emerging areas of application in environment and human health. *Ecosystem Health* 7 (4): 317–325.

Wunder, S. 2007. The efficiency of payments for environmental services in tropical conservation. *Conservation Biology* 21 (1): 48–58.

PART II

PROPERTY RIGHTS AND INSTITUTIONAL ARRANGEMENTS

CHAPTER 5

Why Rights Matter

Lin Crase and Ben Gawne

O ne of the major accomplishments of water reform in Australia has been the establishment of water markets. By international standards, this is a significant policy achievement. Prior to the water reforms of the 1990s, water and land were inseparable—the purchaser of an irrigation farm, for instance, simultaneously secured a right to access a specified volume of water and apply it for a particular purpose. Often the relationship between water and land inputs was also centrally managed. For example, horticulturists were provided a specific amount of water to suit their agricultural pursuits, whereas a different arrangement was in place for those involved in annual cropping, dairy production, and so on. Such arrangements are relatively efficient only so long as the central planner is able to adequately match the various water demands with supply. This task becomes even more problematic when supply becomes constrained or when preferences and demands modify over time.

Centralized planning of water distribution to satisfy multiple users thus requires a substantial amount of information. In the context of Australian water resources, this challenge is exacerbated by the innate variability of water supplies, both temporally and geographically, a point raised in Chapter two. Another major constraint to the planning approach is that it lends itself to considerable political interference. In economic parlance, planning arrangements are susceptible to rent seeking, whereby individuals or groups seek to shape the planning outcomes by expending resources on nonproductive activities, such as bribing officials or using media campaigns to influence public attitudes.

Bennett observes that the worst excesses of rent seeking can be kept in check by a number of forces. These include having a "well educated population, a free press and a responsive democracy" (2010, 3). Combined with a core of knowledge about the values and preferences for different resource allocations, this should give rise to transparency in the decisionmaking process.

Regrettably, these prerequisites are not always easily met, even in well-resourced countries such as Australia. Individuals may be educated but ill informed, the press cannot always be relied on to deliver unbiased views, and the political machinations required for a functioning democracy can prove costly.

It is against this background that markets have long been promoted by standard economic theorists as a means of improving resource allocation. The key advantage of markets is that they turn the rent-seeking game that attends planning exercises into an opportunity for all parties to be better off. Thus, instead of a resource user bemoaning an inadequate allocation by the planner, trade can be used to adjust initial allocations to better suit the needs of the users. In simple terms, trade should see the resource moving from the resource holders who value it least to those who value it most.

An additional and sometimes understated advantage of these arrangements is that values and preferences become more overt. This again limits the rent seeking that can attend planning decisions where resource users deliberately overstate their values as part of the rent-seeking exercise.

Notwithstanding these benefits, there are substantive practical and theoretical challenges to getting markets to work, especially in the context of water. Much of this work pertains to the specification of property rights, along with search, measurement, and enforcement activities to validate exchange. In this chapter, we specifically consider the challenges associated with crafting water property rights and reflect on the Australian experience. A key question for us is, why do tourism and recreation interests make so little use of the water market?

In this chapter, we first briefly consider some of the theoretical nuances of property rights. We then offer a synoptic overview of the rights specification process in various jurisdictions in Australia. Next we turn to some of the dimensions of water for which property rights are not yet fully specified and use this as a basis for speculating on how the different water markets might operate under alternative ownership scenarios. In this discussion, we deal with environmental water management to illustrate the problem because, like tourism and recreation, environmental water uses tend to be nonconsumptive. We then reflect on what this means for tourism and recreational water users more generally before providing some brief concluding remarks.

THEORETICAL NUANCES OF PROPERTY RIGHTS

At the outset, it is important to understand that property rights and ownership are not synonymous. A property right bestows on an individual or group control over a resource. Thus, because rights are ostensibly reciprocal, the holding of a property right amounts to limiting what others may do with that resource (Bromley, 1989). These limits can cover a range of dimensions. For example, Scott (1989) divides rights into six main components: exclusivity of access to enjoying the benefits of the resource; the time span over which benefits can be accrued by the right holder; the transferability of the resource; the divisibility of the resource; the flexibility of the right holder to modify use; and the quality of title, which relates to the capacity to describe the resource, along with sanctions and the responsibilities of right holders.

In addition to understanding that rights are a "bundle" of powers over others, it is also important to appreciate that they are seldom absolute. Rights are likely to be limited or attenuated by superordinate bodies. Regarding the transfer dimension of a water right, for instance, the state may prohibit the movement of water into an area where its application is likely to increase salinity for downstream users. In the context of land, right holders seldom have carte blanche and must comply with a range of restrictions imposed by state or local government bodies. Thus, in addition to having a distributed bundle of rights, the scope of the right holder to act independently with regard to each element of those rights will vary.

Among the most influential works in the discussion of property rights are the seminal papers by Coase (1960). Ronald Coase argues that conflicts over resources and their use could be resolved by greater attention to property rights. At the core of the Coasian argument is the view that harmful or beneficial "spillover effects" should not automatically be considered as an externality, as was the norm in Pigovian economics. For example, if a water user returns pollution to a stream, it is the standard response of Pigovian economists to argue that this is a market failure and requires government intervention to reduce the spillover harm to the wider community. In contrast, the Coasian response would be that establishing rights to pollute the stream could still negate the need for direct intervention by the state. In this case, those who value clean streams could purchase rights from the polluter or vice versa, such that the marginal value of reducing (increasing) pollution was equal across both parties.

The elegance of this solution is that the limits of "planning to reduce pollution" are kept in check. Put differently, once the state is involved in planning to reduce pollution, there is a risk that rent seeking will ensue. Those who value polluting will undoubtedly seek to influence the collective decision by proclaiming massive production losses, while those valuing the reduction of pollution will inevitably seek to overstate the costs. As with all markets, a market in which pollution rights are traded should, prima facie, reveal the value of marginal changes, and those who value rights most will bid them away from those who value them least.

An important caveat to Coasian analysis is that an efficient outcome is contingent on having low transaction costs between those who would trade rights. Again, if we return to the rights to pollute streams, the initial assignment of those rights is largely irrelevant to achieving efficiency, as long as the various parties (pro- and antipolluters) can exchange those rights without incurring other costs. However, market exchange that is costly will reduce the incentives for the various parties to trade, such that all mutually beneficial exchange is not realized. Put differently, if there are significant transaction costs to trade, then the original specification of rights really does matter, and the need exists to closely match the preferences of individuals to achieve a reasonably efficient outcome.

The importance of the initial alignment of rights is made clear if we consider the historical allocation of water and land in Australia to agricultural interests. In many cases, Australia's communal irrigation projects were designed to achieve particular social objectives. More specifically, soldiers returning from war were offered agricultural plots with water access determined by powerful water bureaucracies. The allocation of those land and water resources was determined, at the time, by considering the minimum land and water required to produce standard agricultural outputs and the existing economic value of those outputs. Combining these sources of information, it was then possible to establish what might constitute a viable farming enterprise. In some districts, the planners made serious misjudgments—say, by presuming that the high commodity prices at the time of settlement were sustainable—and thus allocated smaller-than-optimal plots of land. The subsequent hardship endured by farmers in some of these districts is testament to the extent to which rights were originally misaligned. Perhaps not surprisingly, over time farms were progressively aggregated in many of these districts, although not before inflicting substantive disadvantages on some.

In a similar vein, water was allocated to specific pursuits with little account for changing technologies or modifications to the tastes and preferences of end users. The upshot was that in most cases, farmers had either too much or too little water to undertake their preferred activities. For instance, farmers who were originally allocated sufficient water to grow citrus crops using overhead sprinklers soon had excess water rights when they converted to growing grapes under drip irrigation. Moreover, because the nexus between water and land could not be broken, significant inefficiencies attended the specification of rights to water. This was perhaps the greatest motivation for reforming water rights in the irrigation sector, with water trade now commonplace between agriculturalists, at least.

SYNOPTIC OVERVIEW OF RIGHTS DEVELOPMENT AND WATER MARKETS

The core elements of water reform in Australia were described in Chapter 1. In the context of rights, perhaps the most notable starting point was the decision to separate land and water rights in the early reforms of 1994–1995. In essence, each state jurisdiction undertook to break the nexus between land and water and put in place arrangements for the trading of water separately from land. Importantly, the decision was made to define water rights primarily in volumetric terms to facilitate exchange. This is perhaps not surprising, given that the focus at this stage was primarily on consumptive or extractive users of water, a point taken up later in this chapter.

One of the initial and (with the benefit of hindsight) relatively predictable consequences of these arrangements was that underused water rights then assumed a marketable value. In simple terms, water users who had previously used only a portion of their water rights now had a vehicle by which to offer those rights to others in exchange for cash. The upshot was that water extractions increased markedly during this phase, potentially undermining the security of other right holders.

To invoke the property rights parlance, the initial specification of rights was not sufficiently comprehensive to avert spillover effects to other users. In particular, the activation of underutilized water rights had significant consequences for environmental beneficiaries, especially as rights to environmental water were not well depicted in the early reform phase.

A second reform phase commenced in 2004 in the form of the National Water Initiative (NWI). The NWI reemphasized the need to further refine rights in water. Each of the states has pursued the goals established by the NWI with varying degrees of enthusiasm and success. This is particularly notable in the three main irrigation states of New South Wales, Victoria, and South Australia.

In New South Wales, water rights are attenuated by the conditions embodied in water-sharing plans, although these have been broadly suspended by ministerial intervention as a result of severe drought conditions and ongoing policy uncertainty at the federal level. To access water in New South Wales, it is necessary to hold a water access license that represents the right to a clearly defined share of a designated water source. The water access license is separate from any approval rights associated with supply infrastructure and rights associated with use of the water.

To understand the operational aspects of these rights, it is necessary to distinguish between regulated and unregulated water sources. Regulated streams are serviced by dams that can capture water and divert it at times that suit downstream right holders. In contrast, unregulated streams may still be managed by volumetric access rights, but they will be subject to streamflow constraints rather than storage constraints. Put differently, in unregulated streams the rights are less certain, as the right holders have less capacity to manipulate water availability, unless they can store water at sites of their own choosing. In the case of unregulated water sources, a system of rules exists that limits when water can be taken; for example, if streamflows reach a lower threshold, then extraction must cease, or in high-flow events, extraction can occur unabated. The vast majority of trade in water has occurred in regulated streams, especially in the southern portion of the Murray-Darling Basin, and this is the focus of our attention here.

In Victoria, water rights were unbundled in 2007 into three components comprising water shares that represent the maximum volumetric entitlement that can be taken from a water source. The availability of that water is embodied in allocations that are ostensibly the right holder's portion of stored water in a given season or at a point in time in that season. As in New South Wales, use rights are separated from access rights. One nuance of the Victorian communal irrigation areas is the existence of separate rights to delivery infrastructure capability. In simple terms, a farmer can sell his or her water access rights and choose not to use water, but the land owned by the farmer continues to enjoy rights of access to the distribution infrastructure. Unsurprisingly, the right holder also continues to face the responsibilities that attend those rights by payment of charges to the irrigation infrastructure manager.

In South Australia, the single water license that emerged in earlier reforms was converted by the Department of Primary Industry and Resources into four separate components in 2009. First, as in other jurisdictions, a water access license constitutes a perpetual share to a water resource. Second, a water resource works approval relates to infrastructure to intercept water, such as pumps and meters.

Third, a site-specific license attends the use of water. Fourth, allocations are viewed as a separate unbundled right.

The separation of allocations from longer-term water access rights and the fact that the initial assignment of the rights does not precisely meet the needs of right holders has stimulated two main forms of water trade within and between jurisdictions. The trade in water access rights is sometimes referred to as entitlement trade or permanent trade. As these terms suggest, this amounts to the exchange of a right to access a variable supply of water over time. The second form of trade is known as allocation trade or temporary trade. In this instance, the rights being exchanged relate to a designated volume of water that is available during that season.

It is important to appreciate how the allocation procedures operate at a practical level to understand how these markets function and interrelate. In regulated catchments, the dams are generally managed by a centralized dam manager. The role of the dam manager is to measure inflows and calibrate the availability of water against entitlements and other obligations. For example, in addition to accounting for the consumptive entitlements in a dam, the manager may have other rule-based obligations around maintaining minimum transmission flows or securing airspace to limit the prospects of flooding. As a general rule, dam managers also have multiple forms of entitlements to manage. For instance, New South Wales has two classes of entitlement holders, one with a high reliability right and another with a low (termed "general") security of supply. Dam managers have different algorithms for each of these right holders.

In the case of the River Murray, the largest regulated stream in Australia, natural inflows into dams in the headwaters tend to commence in winter and peak in spring. Thus, the dam manager announces an allocation at the start of the season. This allocation indicates to right holders the quantity of water that can be reliably supplied from the available resource at that time. If flows continue into the dam, then subsequent announcements will upgrade the allocation, and in wet seasons, entitlement holders can expect that allocations will reach their full water entitlement. Allocation announcements thus act as information to assist with productive decisions for extractive users.

Overwhelmingly, the larger of the two water markets is the market for allocations. In 2007–2008, 1,237 gigaliters of water allocations were traded in the southern connected Murray-Darling Basin, rising to 1,739 gigaliters in 2008–2009. In contrast, entitlement trade over the same period was 618 gigaliters in 2007–2008 and 1,080 gigaliters in 2008–2009 (NWC 2009). In part, the disparity in traded volumes is due to the relatively modest transaction costs embodied in the allocation market versus the entitlement market.

Brennan (2006) observed that an important element of the allocation market is that it enables an adjustment for extractive users on two fronts. First, the allocation market provides a relatively low-cost means for different water users to redistribute water to account for different opportunity costs. If a perennial horticulturist faces a low allocation with few opportunities for water substitution, he or she can trade with a dairy farmer, who can then use the cash to buy fodder or grain as a substitute for water. Alternatively, an annual farmer might choose to cease production entirely in dry years and sell water allocations. The second important function of the allocation market is that it allows water users with different risk preferences to adjust. Risk-averse farmers facing low allocation announcements can choose to enter the allocation market early in the irrigation season to ensure that they have adequate water on hand for the entire year. By way of contrast, risk seekers might prefer to maintain high levels of production regardless of the initial allocation and take the gamble that inflows will match production or enter the allocation market later in the season. Viewed differently, it can be argued that the water allocation market is used to deal with the mismatch between the way rights are being treated by the dam manager, who may have a very different risk profile than the extractive user, and the preferred right structure of the right holder.

Innovation in water rights in Australia, especially in the form of unbundling, is continuing. In part, this reflects the recognition that the existing rights regimes no longer suffice in all cases. In addition, legislative changes to rights constitute an acknowledgment that markets can be a costly way of correcting an initial misalignment of rights. Notable in this context are the development of carryover rights and the expanding interest in capacity-sharing rights (see, for example, Hughes, 2009).

Carryover rights are an interseasonal transfer right to hold water in regulated catchments. As described earlier, water users can manage the variability of supplies within a given season by accessing the market for allocations or trading away water allocations for cash. Invariably, some right holders will end the irrigation season with excess water. For example, if a risk-averse irrigator purchases additional allocation at the start of the season and inflows realize high allocations by the end of the year, a strong possibility exists that this excess water would be socialized across other users. This will especially be the case if there is no capacity to carry excess water into the following year.

Carryover rights that have developed to date are generally more attenuated than other water rights, in part reflecting the effort to mitigate third-party effects on other dam users. For example, in Victoria, carryover and entitlement water cannot exceed 100% of entitlement before the water moves into a "spillable water account." This account cannot be accessed for trade or use while the dam manager believes there is a chance that the dam will spill; in effect, this water becomes the first to spill in a wet year. In New South Wales, it is not possible to carry over more that 100% of entitlement, and carryover rights have occasionally been extinguished at the whim of government during dry years. The upshot is that carryover rights are less secure than other forms of water rights, and those relying on them for productive purposes face additional risks.

An alternative to carryover is capacity-sharing rights. Two examples exist in Australia in the form of the St. George and MacIntyre-Brooke irrigation areas in Queensland. The basic principle employed in capacity-sharing rights is that individual right holders manage their own storage within the dam and their own risk. This type of right makes allocation announcements redundant, and the dam manager simply measures inflows, credits continuous water accounts, and debits drawdown, subject to transmission and evaporation losses. The advantages of this system are that the risk preferences of the dam manager have no bearing on the way rights are managed, and trade in capacity shares can occur to optimize the benefits of capacity shares across all users. Notwithstanding these advantages, assigning rights in this form also has significant costs, especially where there are existing rights and expectation of water access.

DIMENSIONS OF WATER

As noted in Chapter 2, water has many dimensions beyond its volumetric character. In addition, the previous section highlighted a range of rights that have already been crafted to deal with different elements of water. And yet these rights are far from comprehensive in their coverage. We contend that the heavy focus on specifying the volumetric components of rights is explained by the fact that consumptive users of water have dominated the water debate to date. As tourism and recreation constitute nonconsumptive activities, the rights that would need to be in place to stimulate market participation are broadly absent. Another major nonconsumptive user of water is environmental interests. These interests have expanded their water market activities, but this is not without difficulties. Contrasting the natural characteristics of water is particularly illustrative of some of the deficiencies in existing water rights from the perspective of nonconsumptive water users.

The natural variability of Australia's water resources has often been construed as excessive scarcity (see, e.g., DSE 2005). In reality, it is not scarcity per se that typifies Australian inland waterways, but rather immense variability. Early colonial settlers struggled with this variability, often establishing settlements in relatively wet phases only to face the challenges of unpredictable, severe, and enduring drought after occupation (see Cathcart, 2010). Arguably, the same issue presently confronts the adjustments being sought from the irrigation industry. The growth in extractions from Australian water sources, especially in the Murray-Darling Basin, was most pronounced between the mid-1940s and 1980s, a period of relative wet. In contrast to the data covering a century or so, and particularly the last decade, these extraction levels would now appear to go beyond the sustainable limits of most streams (see, e.g., MDBA 2010).

In addition to interyear variability, the natural condition of many waterways was typified by daily variability and a seasonal pattern that bore little resemblance to the current water regime. The basic response of European settlers to the variability of water supply has been to radically regulate streams. To accomplish this task, large water storage areas have been developed such that water can then be extracted as required. In addition to reducing the volumes of water provided to indigenous flora and fauna, this also amounts to normalizing flows across years, reducing daily variability of flows, and inverting flows so that they no longer peak in the headwaters in winter and spring. Alternatively, flows are now maximized in the hotter, dryer months to suit the needs of extractive uses such as irrigation.

The volumetric changes to streams as a result of European settlement have been the primary focus of concern for environmental interests and scientists who work in this field. Volume has also dominated the area of property rights development.

To give some indication of why volume has captured center stage, it is worth noting that the hydrological modeling undertaken as part of the Murray-Darling Basin planning processes in 2010 showed that the water actually reaching the Murray mouth at the end of the Murray-Darling Basin has, on average, fallen to 41% of the natural outflow, from 12,500 to 5,100 gigaliters per year. One outcome from this has been that the mouth of the River Murray required continuous dredging from 2002 until 2010.

The average reductions in end flows points to the need for additional volumes to be returned to streams to achieve a more sustainable ecological outcome. However, focusing on the volumetric data alone disguises the loss of variability as a consequence of river regulation. This is important because it also brings into question the present specification of rights with such a heavy focus on the volumetric aspects of water.

As we have already noted, water has other important dimensions beyond volume, such as variability of flow between and within particular time frames, including seasonality of flow. In addition, water can vary along quality dimensions, such as the extent of salinity or nutrient loads. These can be particularly important to recreational and tourism interests, with poor water quality resulting in reduced income to providers of water-related recreational activities (see, e.g., Crase and Gillespie, 2008). Water temperature can also play a part, particularly in the context of the provision of water for indigenous fish recruitment and breeding.

To illustrate the problems resulting from the partial specification of water rights, we highlight the challenges arising from the changed environmental demands on water and the current dam management practices described in the previous section. We note that many recreational interests may also coincide with the demands of environmental water managers, a point noted elsewhere in this volume.[1]

Under the current policy arrangements in Australia, a large volume of water is being acquired to address concerns about overallocation and the loss of streamflow, especially in the Murray-Darling Basin. The Productivity Commission (2010) has forecast that under the various federal initiatives, the Commonwealth Environmental Water Holder, the federal agency charged with managing water purchased for the environment, will hold entitlements to about 2,500 gigaliters of average annual flows. Keogh (2010) has suggested that this will represent about 20% of all extractive entitlements held in the Murray-Darling Basin.

To achieve environmental outcomes, the Commonwealth Environmental Water Holder would seek to provide greater opportunities for indigenous species. This can be partially met by simply allowing these volumes to pass through streams, but the other characteristics of these flows are also important. For example, inducing overbank flooding to riparian wetlands and thus replacing some of the missing flow classes removed by river regulation is an important environmental ambition. Similarly, modifying the timing of flows such that river levels rise in winter and spring would be an environmental demand.

However, achieving this type of modification in flow is not well suited to the current market instruments or rights. In a dry year, the manager of environmental water would probably prefer to accentuate the drying cycle and keep any allocations in storage. The drying phase plays an important role in containing invasive species and improving the health of ephemeral wetlands (Gawne and Scholze 2006). In contrast, the environmental water manager would probably prefer to release more water in a wet year to produce overbank flooding of wetlands. To meet this end, it is important to be able to access the maximum amount of water in wet years that has ideally been accrued during dry years. And yet the relative attenuation of carryover rights makes this problematic. In simple terms, because the rights to hold water above entitlement in storage are not separately defined or are defined with significant attenuation, the objective of optimizing the use of this water for environmental ends becomes more costly.

An alternative to using carryover rights in this manner would be to use the allocation market to supplement riverflows during wet years. This would also have the advantage of accessing relatively low-cost allocations, as they can logically be expected to be cheaper in wet years and the cost could be offset by selling allocations in dry years. However, practical and political challenges exist on this front. More specifically, it seems clear that the environmental water holder is unlikely to be permitted to actively trade water, at least at this stage.

A similar limitation to the structure of water rights emerges in the context of the seasonal distribution of flows. Because the environmental water holder has purchased entitlements, and releases are subject to the standard dam management rules, it is difficult and costly to achieve earlier flows than what has been the convention. We noted earlier that dam management is subject to allocation announcements that start conservatively at the beginning of the season and rise with inflows. Ideally, an environmental water manager would prefer to see flows released earlier in the irrigation season, around winter and spring. But because allocations are counted from the commencement of the season and carryover is attenuated, the holder of environmental water cannot easily achieve this end. The market for allocations could be used in this context, but water allocations tend to be more expensive at the start of the season, again placing the environmental water manager at a disadvantage compared to those with an interest in summer flows. Clearly, those with an interest in the timing and variability dimensions of water seem to have less capacity than those with an interest in volumes of water.

CONCLUSIONS

At the crux of the challenges facing the holder of environmental water is the fact that the rights to the water entitlements on hand do not capture their relatively recent interests. In essence, the rights of a party with an interest in the nonconsumptive dimensions of water appear weaker than the rights of those with an interest in consumptive activities. Tellingly, this seems at odds with the changing economic fortunes of the various sectors in Australia. For example, numerous studies attest that the value of environmental amenity is increasing in countries like Australia, in part as a result of increasing wealth. Similarly, the value of tourism and recreation continues to increase. Simultaneously, the sector with the most interest in the extractive use of water, irrigated agriculture, continues to suffer from diminishing terms of trade. Prima facie, this should set the scene for significant trade in water rights across the various sectors, even if rights were slightly misspecified to begin with. As we have noted, it would be possible to use the market to overcome some of this misalignment of rights, but that does not come without cost. Moreover, it will be interesting to observe whether a readjustment of rights is forthcoming, given that the rights to such a large quantity of water now held by environmental interests are potentially misspecified.

Other chapters in this volume highlight the present mechanisms by which the tourist sector's interests in water are expressed. In that context, we have two important observations. First, those with an interest in tourism or recreation presently endeavor to influence water allocation decisions via the political domain and through the shaping of the rules and constraints around marketable water rights. As a general rule, entering the market to purchase rights is not the modus operandi to express tourism or recreational interests in water. Second, because tourism and recreational interests in water vary markedly, the outcomes sought by these interests also vary. Some interests align with those with a preference for ecological outcomes, whereas others are more conducive to the conventional water management regime that aligns with irrigation. Arguably, other tourism and recreation interests are independent of both agricultural and environmental water users. In any case, the question remains as to why such interests do not emerge in the existing water markets.

As noted earlier in this chapter, Australia has two main markets for water: a market for allocations and another for entitlements. In both cases, these represent volumetric amounts of water, with flow and storage management largely attenuated by the way dams are centrally managed. Recreational users of water generally have little interest in volumes of water per se. Rather, their interest lies in the availability of flows and stored water to undertake their leisure pursuits. For example, the owner of a houseboat has little interest in a volume of water—it is simply not feasible to isolate a small volume and secure it under the boat. The houseboat owner would prefer that flow and storage be managed to suit his or her interests. Thus, securing the interests of the houseboat owner via the present market would require an excessive purchase of water, beyond that interest. In effect, the houseboat owner would need to buy the water in the dam. This is analogous to asking a commuter on a coach to buy the bus to get it to run at a different time of day.

Engagement of tourism and recreation interests in water markets is feasible, but for want of adequately defined rights. This means being able to separately specify and trade some of the nonconsumptive elements of water, especially storage and flow dimensions. In the previous section, we noted that pressure has been increasing to consider rights and markets to allow trade around the seasonal availability of water in streams and storage areas. This is being driven by the environmental interest in the nonconsumptive element of water and how this can be manipulated to achieve ecological gains. Further development on this front offers much promise for the tourism and recreation sector. In essence, the contribution of recreation and tourism in such markets may be sufficient at the margin to support a range of adjustments to water management in this country.

For example, a farmer who agrees to take his or her water at a time that suits other interests could be rewarded by trading away control over the delivery time. This would have obvious advantages to both recreational users and environmental interests who could genuinely participate in such markets to secure the flow and storage outcomes that suit.

In sum, the initial specification of rights in water does matter, especially when adjustment via the market becomes costly. The evolving interests in water and changing preferences for leisure and environmental amenity are important in this context. The evolution of Australian water markets offers valuable lessons about crafting rights in water and shows how some interests will inevitably be advantaged. However, it is also important to realize that rent seeking will inevitably ensue whenever a planning approach is adopted. In that regard, markets offer great promise for easing the conflict over water resources. Nevertheless, specifying rights that represent a range of interests, including those of tourism, should be a policy priority.

NOTE

1. This is not to say that all recreational and tourism interests align with environmental interests. For instance, those seeking to retain water in storage over summer for boating may have a different interest in water releases from dams than those who enjoy recreational fishing for indigenous species.

REFERENCES

Bennett, J. 2010. Informing tough trade-offs. paper presented at *Water Policy in the Murray-Darling Basin: Have We Finally Got It Right?*, October 21–22, 2010, University of Queensland, Brisbane.

Brennan, D. 2006. Water policy reform in Australia: Lessons from the Victorian seasonal water market. *Australian Journal of Agricultural and Resource Economics* 50 (4):403–423.

Bromley, D. 1989. *Economic Interests in Institutions: The Conceptual Foundation of Public Policy.* New York: Basil Blackwell.

Cathcart, M. 2010. *The Water Dreamers: The Remarkable History of Our Dry Continent.* Melbourne: Text Publishing Company.

Coase, R. 1960. The problem of social cost. *Journal of Law and Economics* 3 (1):1–44.

Crase, L., and R. Gillespie. 2008. The impact of water quality and water level on the recreation values of Lake Hume. *Australasian Journal of Environmental Management* 15 (1):21–29.

DSE (Department of Sustainability and Environment). 2005. *Securing Our Water Future Together: Key Concepts Explained.* Melbourne: DSE. Fact Sheet.

Gawne, B., and O. Scholz. 2006. Synthesis of a new conceptual model to facilitate management of ephemera deflation basin lakes. *Lakes and Reservoirs: Research and Management* 11:177–188.

Hughes, N. 2009. Management of irrigation water storages: Carryover rights and capacity sharing. paper presented at *Australian Agricultural and Resource Economics Society Annual Conference*, February 10–13, 2009, Cairns.

Keogh, M. 2010. Background. In *Making Decisions about Environmental Water Allocations—Research Report*, edited by M. Keogh and G. Potard. Surrey Hills, Australian Farm Institute, 3–5.

MDBA (Murray-Darling Basin Authority). 2010. *Guide to the Proposed Basin Plan.* Canberra: Murray-Darling Basin Authority.

NWC (National Water Commission). 2009. *Australian Water Markets Report 2008–09.* Canberra: National Water Commission.

Productivity Commission. 2010. *Market Mechanisms for Recovering Water in the Murray-Darling Basin: Final Research Report.* Melbourne: Productivity Commission.

Scott, A. 1989. Conceptual origins of rights based fishing. In *Rights Based Fishing*, edited by P. Neher and R. Arnason, N. Mollet. Dordrecht: Kluwer Academic, 11–38.

CHAPTER 6

Institutional Considerations for Collaborative Behavior in Tourism and Recreation

Brian Dollery and Sue O'Keefe

*I*n historical terms, Australian regional water policy has hinged on the assumption that fresh water was predominantly an input factor into agricultural and industrial production processes in the hinterland and consequently an important determinant of regional development. For this reason, regional water policy was primarily seen as a tool for stimulating regional economic development, and water rights were thus allocated in a "top-down" fashion in line with this goal. Over the first century of federation, water underpinned almost all efforts at inland regional agriculturally based growth policy, such as agrarian population redistribution to country areas from the coastal cities, soldier resettlement schemes, and the like. However, over the past three decades, a growing awareness of other imperatives, not least the prevention of environmental degradation in inland Australia, has become much more significant in the shaping of nonmetropolitan water policy.

As discussed earlier in this volume, in the evolution of Australian water policy, the so-called "development" hypothesis has been succeeded by a new "management" philosophy focused on dealing with a much more mature water economy marked by an inelastic supply of existing fresh water. Recognition of water scarcity and conflicting demands for water from a wide array of users has obliged water policymakers to embrace multifarious water policy objectives, including economic efficiency, sustainable development, and ecological sustainability. A significant consequence of this change in thinking has been the fact that water policy formulation, implementation, and management have become not only much more complicated, but also more controversial as policymakers seek to balance incompatible demands from different water users.

In general, most scholarship has concentrated on the problem of how to reconcile the conflicting interests of consumptive users, such as industrialized large-scale agriculture, with the need to restore nonconsumptive water for environmental ends.

However, this broad characterization of the trade-offs involved in water usage has neglected important user subgroups that have vital interests in the allocation of water. For example, researchers have shown scant regard for recreation and tourism as users of water, despite the growing economic significance of these activities.

This neglect is unfortunate for several reasons. In the first place, in many respects the recreational use of water by both tourists and the local community alike contains in microcosm the same kinds of dilemmas implicit in the development-versus-conservation bifurcation. After all, the recreational use of water for tourism simultaneously has ecological sustainability at its core, as in the case of environmental tourism, and consumptive water application, as in the direct demand for water by tourists and other recreational users for water. Second, because these tensions can be manifested in a single industry, such as regional tourism, scholars can examine them without excessive interindustry institutional complications. Third, the recreational use of water is growing in importance in regional Australia as the composition of economic activity in these areas shifts away from its traditional reliance on agriculture and agricultural service industries, and it thus can no longer be simply ignored. Finally, inclusion of recreation and tourism as significant elements in the water allocation process complicates an already difficult terrain for public policy and thereby obliges policymakers to venture even more deeply into the real world of hard choices. For these reasons, Chapter 5 considered the academic literature on public policymaking, notably collaboration in water policy, water administration, and water usage, drawing on examples of the recreational and tourist consumptive and nonconsumptive demands for water.

As in many other countries, the complexities of water policymaking in contemporary Australia have served to underline the systemic weaknesses of centralized top-down decisionmaking of the kind that led to the collapse of socialism in the 20th century. Moreover, in the literature on natural resource management, centralized "command-and-control" methods of management have been attacked for exacerbating the vulnerability of resource-dependent communities and damaging ecological sustainability (Colfer 2005; Zerner 2000). Recognition of these shortcomings in centralized decisionmaking systems has fostered a new paradigm with three main foundations. First, broad-based participation is essential in the development of policy to ensure that local knowledge and interests are taken into account. Second, a need exists to embrace "knowledge, learning and the social sources of adaptability, renewal and transformation" (Armitage et al. 2007, 2). Finally, change and uncertainty are inherent in socioecological systems. An important consequence has been the development of collaborative approaches to natural resource management as the best methods of tackling complex and contentious natural resource problems.

Numerous terms have been coined to encapsulate this decentralized collaborative approach to decisionmaking in natural resources management (see, e.g., Conley and Moote 2003). For instance, collaborative methodologies have been described as "resource partnerships" (Williams and Ellefson 1997), "consensus groups" (Innes 1999), "community-based collaboratives" (Moote et al. 2000), and "alternative problem-solving efforts" (Kenney and Lord 1999). In a similar vein, collaborative natural resource management has been phrased as "watershed management"

(NRLC 1996), "collaborative conservation" (Cestero 1999), "community-based conservation" (Western and Wright 1994), "community-based ecosystem management" (Gray et al. 2001), "grass-roots ecosystem management" (Weber 2000), "integrated environmental management" (Margerum 1999), and "community-based environmental protection" (EPA 1997). In addition, particular conceptual models have been created, such as "coordinated resource management" (Cleary and Phillippi 1993) and "collaborative learning" (Daniels and Walker 2000). While there are nuanced differences between many of these expressions, they undoubtedly share much in terms of common characteristics.

This chapter starts by examining the transformation in policymaking that has occurred in all areas of public policy, including natural resource management policy, contingent on the recognition that central government control could not deliver satisfactory results, by considering the revolution in economic thinking as scholars came to understand the pervasive nature of government failure. We then build on this discussion by outlining the main elements of decentralized resource governance that flowed from these insights. This is followed by a synoptic review of decentralized environmental policy instruments suitable for decentralized governance structures. Next we examine decentralized resource governance and collaborative hybrid resource management models. The chapter ends with some brief considerations on the best way forward in collaborative decentralized resource management.

GOVERNMENT FAILURE AND PUBLIC POLICYMAKING

Environmental governance has experienced an immense shift in emphasis over the past few decades. For most of its history, since at least the 19th century, natural resource management has been dominated by a desire to use centralized bureaucratic agencies to deal with questions of environmental concern, such as biodiversity, deforestation, desertification, and the exhaustive exploitation of natural resources, through various kinds of direct and coercive state control. In their description of this long-held view, Lemos and Agrawal point out that "state bureaucratic authority appeared to many policy makers and academic observers as the appropriate means to address the externalities associated with the use of environmental resources", with "centralized interventions" thus "essential to redress resulting market failures" (2006, 303). However, this faith in the capacity of the state to solve societal problems, including environmental issues, has been eroded over the past three decades for a host of reasons, not least the observed inability of government to generate optimal solutions to economic and social problems. Because this disillusionment with centralized governmental decision-making has played a pivotal role in public policymaking in the natural resource management arena, it is worth considering in more detail.

The "public interest" approach, which underpinned this central-control model of natural resource management, was based on the theory of market failure and its implicit conception of an idealized state. It was founded on at least three untenable assumptions. In the first place, it presumed that policymakers could accurately determine the extent of market failure. In the realm of natural resource management and water policy, this tacit claim mostly surrounded accurately estimating the size of negative externalities, a notoriously difficult empirical problem. Second, it presupposed that central government agencies possessed the ability to intervene efficiently to correct perceived instances of market failure. For natural resource management and Australian freshwater allocation policy, this implied that policymakers could design optimal regulation, taxation, and subsidy policy tools that would reduce rather than increase the welfare loss in question, which is a brave assumption. Third, it accepted that policymakers would apply public policy in an altruistic manner congruent with the "public interest," in the face of evidence to the contrary, as policymakers were "players" in the process and not merely neutral referees. Even the most cursory glance at the highly politicized nature of contemporary Australian water policy would undermine this proposition.

By the mid-1960s, numerous scholars began to question the public interest approach, with its underlying market failure paradigm and heroic presumption of a benevolent and omnipotent state. Four major areas of criticism emerged, as Wallis and Dollery (1999) note:

- Critics attacked the assumption that the state could somehow accurately determine the magnitude of welfare losses flowing from market failure, and then implement optimal policy measures. For example, in *Law, Legislation and Liberty*, Hayek denounced this claim as a "synoptic delusion" or "the fiction that all the relevant facts are known to some one mind, and that it is possible to construct from this knowledge of the particulars a desirable social order" (1973, 14). It was argued that given our limited understanding of economic processes, including natural resource science and technology, it is highly unlikely that state agencies could possess adequate knowledge to intervene rationally.
- Opponents questioned the ability of governments to intervene effectively in the public interest. They identified a number of factors that inhibited the capacity of the central state to respond fully and efficiently to the needs of the citizenry.
- Observers rejected the assumption of altruistic behavior underlying the public interest approach in favor of a self-interested model based on the standard *Homo economicus* postulate of economic theory conventionally applied to consumers and producers. Downs put the argument as follows: "Even if social welfare could be defined and methods of maximizing it could be agreed upon, what reason is there to believe that the men who run the government would be motivated to maximize it?" (1957, 136).
- Lipsey and Lancaster (1956) developed the theory of the "second best," which undermined the desirability of government intervention aimed at generating optimal economic conditions, even if policymakers knew the extent of market failure, intervened efficiently, and framed policy in an altruistic manner. The primary reason for this lay in the fact that the second-best paradigm demonstrated that if market failure was present in one sector of the economy, then a higher level of social welfare may be attained by deliberately violating efficiency conditions in other sectors, rather than by intervening to restore economic efficiency in the initial case of market failure.

This paradigm shift in perceptions on the nature of state intervention has had dramatic implications regarding economic, environmental, and social policy. The earlier view that market failure necessitated public policy intervention aimed at creating economic efficiency, and thus the actions of policymakers could be explained as a benevolent attempt at generating the optimal conditions required for maximizing social welfare, no longer held sway. It was replaced by a new, more skeptical view that stressed the problems associated with government intervention and the self-interested motivation behind such intervention. In this policy paradigm, the apparent inability of public policy to achieve socially optimal outcomes was termed "government failure." It held that the costs attendant on government failure should be set against the purported benefits of intervention designed to ameliorate market failure.

This insight has spawned a voluminous literature on government failure with two main strands: the positive theory of government failure and a normative government failure paradigm (Wallis and Dollery 1999). In general, this literature has rejected the "benevolent state" conception of the market failure paradigm by assuming that politicians and public servants act in terms of the *Homo economicus* assumption in common with producers and consumers. Given the momentous changes that have taken place in environmental policymaking, it is worth examining this literature in more detail.

In terms of the positive theory of government failure, the earliest modern approach to this phenomenon was the "capture" theory of regulation initially developed by Stigler (1971), who argued that industry regulation is "captured" by the industry under regulation and "designed and operated primarily for its benefit" (1975, 114). Peltzman (1976) extended Stigler's (1971) model by arguing that regulation is supplied by politicians seeking to maximize votes by responding to interest groups representing consumers and producers. In terms of environmental regulation and the Australian water debate, the main thrust of this theoretical perspective was that the regulation process can be captured by industrial interests and manipulated to operate in their favor.

However, the most significant modern approach to government failure is provided by public choice theory. In essence, public choice theory applies the standard *Homo economicus* to nonmarket or political processes underlying policy formulation and implementation. This methodology has given rise to various typologies of government failure, which can be useful in the analysis of real-world policy debates, including environmental policy debates involving water allocation. One of the earliest contributions was advanced by O'Dowd (1978), who argued that all forms of government failure can be classified in terms of three generic types: "inherent impossibilities," "political failures," and "bureaucratic failures." O'Dowd described these categories as follows:

> The first type covers the cases where a government attempts to do something which simply cannot be done; the second, where although what is attempted is theoretically possible, the political constraints under which the government operates make it impossible in practice that they should follow the necessary policies with the necessary degree of consistency and persistence to achieve

their stated aim, [whereas] the third type covers the cases where although the political heads of the government are capable of both forming and persisting with the genuine intention of carrying out a policy, the administrative machinery at their disposal is fundamentally incapable of implementing it in accordance with their intentions. (1978, 360)

More recently, Dollery and Wallis have proposed a closely related tripartite taxonomy, under which system "legislative failure" refers to the economic inefficiency that derives from the "excessive provision of public goods as politicians pursue strategies designed to maximise their chances of re-election rather than policies which would further the common good." Even if socially beneficial policies were enacted, "bureaucratic failure" will inhibit these policies from being effectively implemented because "public servants lack sufficient incentives to carry out policies efficiently" (1997, 360). Finally, since government intervention almost always involves wealth transfers, rent seeking will inevitably accompany such intervention, with socially harmful consequences.

Weisbrod developed the most comprehensive typology of government failure, with a quadrilateral taxonomy: "legislative failure," with the same meaning as in Dollery and Wallis (1997); "administrative failure," based on the proposition that the "administration of any law inevitably requires discretion" and the "combination of information and incentives acts to affect the manner in which the discretion is exercised"; "judicial failure," when the legal system fails to deliver judicially optimal outcomes; and "enforcement failure," defined as the suboptimal "enforcement and non-enforcement of judicial, legislative, or administrative directives" (1978, 36–39).

In contrast to this positive theorizing, scholars have made various attempts to construct a normative theory of government failure, most notably Wolf (1979a, 1979b, 1983, 1987, 1989), Le Grand (1991), and Vining and Weimer (1991). In the first place, Wolf assembled a theoretical framework as a conceptual analogue of the market failure paradigm to "redress the asymmetry in the standard economic treatment of the shortcomings to markets and governments by developing and applying a theory of 'non-market'—that is government failure—so that the comparison between markets and governments can be made more systematically, and choice between them arrived at more intelligently" (1987, 43). His model thus mirrored the theory of market failure by imputing various kinds of nonmarket failure to idiosyncrasies in underlying "demand" and "supply" conditions on the premise that "just as some types of incentive encourage market failure, so too incentives influencing particular non-market organizations may lead to behavior and outcomes that diverge from ones that are socially preferable, according to the same criteria of preferability as those for market efficiency and distributional equity" (Wolf 1979b, 117-118).

Following this procedure, Wolf identified four generic types of "non-market failure" (1979b, 117–118):

• "Internalities and private goals," which refer to the intraorganizational allocation and evaluation procedures of public agencies or their "internal version of the price system." Unlike the "internal standards" of market organizations, normally strongly linked to the "external price system," nonmarket organizations often have internalities largely unrelated to optimal performance. This could mean that the actual behavior of a public agency may diverge from its intended or ideal role. For example, in environmental policymaking and enforcement, field inspectors may be given productivity targets measured in terms of the number of visits rather than the substance of those visits (i.e., a bias toward process rather than outcome).

- "Redundant and rising costs," which represent another kind of nonmarket failure, because while market processes impose a relationship between production costs and output prices, this relationship is generally absent in nonmarket activity as revenues derive from nonmarket sources, such as government tax income. Accordingly, "where the revenues that sustain an activity are unrelated to the costs of producing it, more resources may be used than necessary to produce a given output, or more of the non-market activity may be provided than is warranted by the original market-failure reason for undertaking it in the first place" (Wolf 1989, 63). In environmental policy implementation, in common with other public sector activities, this may mean agency resources are not used efficiently.

- "Derived externalities," which are the unintended and unanticipated consequences of government intervention designed to ameliorate market failure. Thus, in common externalities generated in market relationships, these represent costs and benefits not considered by economic agents, so derived externalities in the nonmarket sphere are "side effects that are not realized by the agency responsible for creating them, and hence do not affect the agency's calculations or behavior." In environmental policymaking and enforcement, unintended consequences of this kind often arise, such as the rise of the malarial mosquito problem after the banning of DDT in many parts of the developing world.

- Adverse distributional consequences, in which case it is argued that nonmarket inequities characteristically occur in terms of power and privilege, whereas distributional market failure typically appears in income and wealth differences. Recent controversy in New South Wales land use policy has seen instances of this kind, where farmers are arbitrarily forced by regulation to cease using their land for productive purposes, thereby bankrupting them.

Building on the work of Wolf's model, Le Grand developed an alternative model of market failure that was "analytically more precise and more comprehensive" than the conventional theory (1991, 424). Le Grand's model of government failure consisted of a tripartite classification of government intervention in a market economy, evaluated in terms of two measures of economic efficiency and an equity criterion. Le Grand argued that government can involve itself in economic, environmental, and social activity in three ways: through provision, taxation or subsidy, and regulation.

In common with Le Grand, Vining and Weimer (1991) proposed a new normative model of government failure based on a critique of the Wolf approach. In government production, Vining and Weimer postulated "contestability of supply," which dealt with the competition faced by a public agency for its output. Given difficulties in monitoring the supply of complex services, they contended that "trust" becomes a critical ingredient of sound public policymaking. Vining and Weimer set out their argument as follows: "When the risk of opportunism, determined by the cost of non-compliance and the opportunity for non-compliance in contracting, is high, trust is an important specific asset to the production of the good" (1991, 6–7). Under these circumstances, the supply of the good is unlikely to be contestable, and government production is an appropriate organizational arrangement when the risk of opportunism is high. A second attribute of the contestability of government production of public goods resides in contestability of ownership or "the credibility of the threat of transfer of ownership of the organization" (Vining and Weimer 1991, 6–7). From the perspective of positive economics, where both supply and ownership are highly contestable, economic efficiency results and prices will reflect marginal cost. Contracting will be the most socially efficient mode of provision. The opposite conclusion applies where contestability is low; in those instances, government production will be more socially efficient.

From this brief synoptic review of the literature, it is evident that a highly evolved literature on government failure exists alongside the longstanding market failure paradigm. In the realm of public policymaking, including natural resource management generally and Australian water policy in particular, a holistic conception of public policy should embrace both of these paradigms, as they oblige policymakers to consider the comparative advantages of market and nonmarket mechanisms in securing economic, environmental, and social objectives (Wolf 1989). Notwithstanding these insights, however, once a decision has been made to employ market or nonmarket instruments, only limited light is shed on the problem of establishing which market or nonmarket instrument or instruments would be the most efficient method of achieving a policy goal. For example, while a decision to employ a decentralized approach to water management follows directly from the policy prescriptions of the government failure literature, it still requires a great deal of further reflection on what kinds of public, private, and nonprofit participants should be involved and how decisionmaking should occur.

DECENTRALIZATION AND RESOURCE MANAGEMENT

The preceding discussion makes it clear that a tectonic shift has occurred in the conceptual basis for public policymaking in democratic market economies over the past 40 years. Recognition that highly centralized command-and-control models of governance centered in large national government bureaucracies are a virtual recipe for extensive government failure in natural resource management, especially where the resource in question has a significant spatial dimension and where substantial regional variations are evident, has engendered a great deal of scholarship on decentralized governance. In particular, work on common property regimes and alternative political structures has focused on the capacity of small local communities to manage natural resources. This has served to furnish the intellectual foundations for the comanagement of natural resources, community-based natural resource management, and environmental policy decentralization (Ostrom 1990).

Lemos and Agrawal have argued that this corpus of work accomplished this transformation "by demonstrating that forms of effective environmental governance are not exhausted by terms such as 'state' and 'free market institutions' and that users of resources are often able to self-organize and govern them." Moreover, "by identifying literally thousands of independent instances of enduring governance of resources and at the same time highlighting arenas in which external support can improve local governance processes, scholars of common property and political ecology have helped prepare the ground for decentralized environmental governance." The result has been that "since the mid-1980s, decentralization of authority to govern renewable resources, such as forests, irrigation systems, and inland fisheries, has gathered steam" until it has become "a characteristic feature of late twentieth and early twenty-first century governance of renewable resources" (2006, 303).

Although it is not commonly recognized, the case for the decentralization of environmental governance has drawn heavily on earlier work on the economics of fiscal federalism by Oates (1972). While the literature on government failure has served an invaluable function in alerting public policymakers to the fact that they must choose between the necessarily "imperfect alternatives" of markets and governments (Wolf 1989), it has provided much less insight into how decisions should be made among different tiers of government in a federal system, although an embryonic strand of this literature has focused on competition among governments in a multilevel state (Breton 1995). However, some light on this vexing question can be shed by work on decentralization and efficiency in multi-tiered systems of government (see, e.g., Ahmad and Brosio 2009; Oates 1999).

In the fiscal federalism literature, invoking the principle of subsidiarity, it is argued that fiscal decentralization through the assignment of taxation and expenditure functions to lower levels of government can enhance the efficiency of public policy formulation and implementation through two main mechanisms. First, on the demand side, the "preference-matching hypothesis" holds that decentralization can improve economic efficiency, because lower levels of government may provide public goods and services better matched to local preferences. Second, on the supply side, decentralization may improve the provision of public services for several reasons: it gives voters greater control over politicians and bureaucrats, thereby reducing rent seeking; it encourages "yardstick competition" between subnational governments with efficiency dividends; it inhibits interest group lobbying, thus decreasing policy distortions and wasteful expenditure; and it allows subnational public agencies with superior local knowledge greater control over local expenditure. It is obvious that this line of thought has substantial relevance to the debate on natural resource governance.

Following these insights from the theory of fiscal federalism, in the realm of environmental governance, advocacy of decentralized governance has generally invoked three propositions (Hutchcroft 2001). In the first place, it is argued that decentralized governance can yield greater efficiency because of competition between subnational units of government and other participating public entities, nonprofit organizations, and community groups. In terms of the Australian water policy debate, this points to the manner in which different solutions to water allocation problems can arise if different groups of stakeholders are allowed to develop their own solutions to local water allocation problems. In effect, a real-world "laboratory" is established in which different groups strive independently to find solutions to water problems. Second, it is held that decentralized governance can bring decisionmaking nearer to those most closely affected, thereby promoting higher participation and greater accountability. Finally, superior local knowledge can become available to decisionmakers concerning the place-specific natural resources.

In environmental policymaking, disaffection with top-down, command-and-control, centralized methods of decisionmaking and the search for improved decentralized approaches to natural resource problems has had two main effects. First, environmental policy design has sought to incorporate incentive mechanisms for modifying individual behavior to align it with desired outcomes. In practice, this has involved the adoption of market processes and incentives. Second, natural resource policy has focused on the creation of new governance structures that are better suited to decentralized environmental governance.

DECENTRALIZATION AND ENVIRONMENTAL INCENTIVE INSTRUMENTS

A key ingredient in the evolution of thought in environmental policy design has been the embrace of incentive mechanisms typically employed in exchange or market relationships. The basic idea underlying the adoption of this approach is to harness individual incentives to change individual behavior by modifying the costs and benefits associated with particular environmental strategies. To recalibrate the calculus individuals perform in deciding how to tackle tasks with environmental consequences, policymakers invoke incentives to try to ensure improved outcomes.

At the conceptual level, it is possible to identify three generic types of environmental management tools:

- direct regulation, typically in the form of technological restrictions, such as meeting specified purity standards, mandatory abatement control, or appropriate waste management;
- cooperative institutions that share information among regulators, polluters, and victims so that the extent of externalities are known and their consequences appreciated; and
- market incentives that increase the cost of environmental degradation and provide benefits for its prevention.

The third of these generic environmental management tools, market incentives, can in turn be broken down into three broad categories:

- price rationing, often in the form of emissions or effluent charges;
- quantity rationing, usually in the form of permits and tradable emissions instruments; and
- liability rules, which impose benchmark standards and penalize violations of these benchmarks, such as with noncompliance fees, deposit-refund schemes, or performance bonds.

In essence, these market instruments have three main advantages over direct regulation and other methods of environmental control. In the first place, market tools, such as charges, taxes, and tradable instruments, allow producers and consumers to reduce harmful activity in a least-cost manner, thereby achieving the "optimal" level of pollution. Second, market-based incentives provide an ongoing incentive to continually improve products and processes. Finally, in the context of public finance, they enable governments to raise revenue through environmental policy.

Against this theoretical background, it is possible to identify a wide range of instruments that have been developed to assist decentralized governance, including ecological taxes and subsidies, often an admixture of public regulation and individual incentives, tradable permits, voluntary agreements, ecological certification programs, and product information requirements such as "eco-labelling" (Tews et al. 2003). In contrast to central command-and-control mechanisms, these policy instruments all rest on the *Homo economicus* assumption of individual motivation.

At the most decentralized level of government, which is typically local government in most countries, environmental taxes and charges are usually levied as a form of price rationing of environmentally undesirable conduct. In general, three types of taxes and charges characteristically dominate this category of environmental policy instruments: emissions charges, ambient charges, and product charges. Emissions charges represent fees levied on the discharge of pollutants into the environment and are intended to improve the quality of the local environment. At the local-government level, in the Australian Westminster model, emissions charges are most commonly levied on domestic and industrial waste. The revenue raised through these charges can be used to defray the costs of storage in municipal dumps and the like, as well as treatment of the waste to render it less harmful. In general, emissions charges of this kind are justified on grounds that domestic and industrial waste must be treated and stored to meet specified environmental standards, and that user charges are the most appropriate and equitable way of financing the costs involved.

Ambient charges are used where emissions charges cannot be employed because of moral hazard due to asymmetric information. For instance, in a given area, the regulatory authority might be able to acquire accurate information only on the aggregate concentration of an air pollutant; it thus cannot feasibly tell how much each household or firm has contributed to the air pollutant. It can then impose an ambient charge scheme that either rewards or punishes all households and firms involved equally. An example of this would be the Australian approach to charges for the smoke caused by domestic wood fires used for heating homes. Because it is too expensive to monitor each household chimney individually, and only aggregate levels of wood smoke pollution are known to the regulating authority, emissions charges cannot be used. Thus in Australia, local councils sometimes provide subsidies to households to remove old wood heaters and replace them with more efficient modern wood heaters or even gas heating systems. However, the legal complexities surrounding ambient charge schemes in the local milieu, as well as the political obstacles typically involved in implementing ambient schemes, mean that they are not frequently used.

Product charges represent fees or taxes levied on inputs or outputs hazardous to people in the local environment. Product charges can be levied at any stage of the product cycle, including production, use, and disposal. They can also be levied on particular characteristics of pollutants, such as the persistence of a pollutant over time. Thus many types and variations of product charges exist. For instance, plastic bags are taxed in South African supermarkets to discourage their use by consumers, who must bear the additional charge.

Tradable permits represent an alternative decentralized approach to the problem of negative environmental externalities. In common with environmental taxes and charges, tradable permits attempt to internalize externalities into the behavior of consumers and producers by forcing them to take into account the full costs of their conduct. The conceptual foundations of tradable permits rest on the proposition, developed by Coase (1960), that the economic efficiency of resource allocation can be improved through market exchange only if legal and institutional arrangements allow resources to flow to their highest market use value. By selling such permits, and rationing the number available for trade, the regulating authority can not only raise revenue, but also limit environmental damage. In Australia, tradable water permits have become a critical method of allocating water among competing users.

Voluntary environmental agreements typically involve large corporations that undertake to meet voluntarily imposed environmental targets. As this has often been done to forestall legal regulation, it has been argued that these kinds of agreement will be effective only if the threat of legal regulation is maintained (Segerson and Miceli 1998). Ecolabeling and environmental certification schemes represent subgroups of the more generic category of voluntary agreements. Under these programs, producers typically consent to specified environmental standards and relay these agreements to the public in terms of "environmentally friendly behavior" in marketing campaigns. In practice, primary sector corporations, such as coffee, energy, and seafood producers, frequently adopt ecolabeling and certification schemes.

DECENTRALIZATION AND COLLABORATIVE MANAGEMENT

In addition to the incentive-based policy instruments developed to aid decentralized environmental policymaking, it is also possible to identify a host of new institutional arrangements that have grown up to foster decentralized environmental governance. In contrast to past experience in natural resource management, contemporary development in this area has largely cultivated hybrid governance models that cross the traditional divides among community, market, and state. A useful illustrative approach to appreciating the complexities of these new models has been provided by Lemos and Agrawal (2006), which is reproduced here in Figure 6.1. This figure depicts a schematic method of classifying strategies of environmental governance combining community, market, and state.

It is evident from Figure 6.1 that three hybrid decentralized governance models for environmental policy and natural resource management have emerged:

- *comanagement*, involving relationships between public agencies and local or regional communities;
- *public-private partnerships*, encompassing relationships between government departments and private firms; and
- *private-social partnerships*, constituting relationships between private entities and local or regional communities.

An additional genre of models, not shown in Figure 6.1, comprises partnerships among the three core social categories: community, market, and state.

Attempts by economists to differentiate among alternative institutional solutions to economic, environmental, and social problems often take as their point of departure the emphasis given by seminal thinkers in the New Institutional

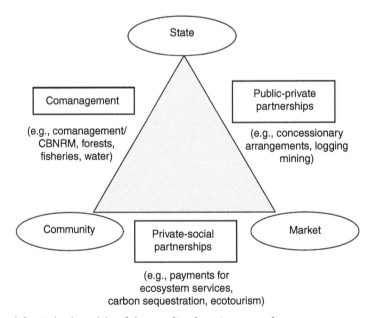

Figure 6.1 *Hybrid models of decentralized environmental governance*

Source: Lemos and Agrawal (2006, 310, Figure 1)

Economics tradition, such as Coase (1937) and Williamson (1985), to markets and hierarchies as distinct governance mechanisms associated with specific types of transaction costs. Subsequent developments in this tradition have added a third category to this scheme. Different triads of terms have thus emerged: markets, hierarchies, and networks (Thompson et al. 1991); community, market, and state (Streek and Schmitter 1985); markets, bureaucracies, and clans (Ouchi 1991); price, authority, and trust (Bradrach and Eccles 1991); and markets, politics, and solidarity (Mayntz 1993). All of these hark back to Boulding's (1978) distinction among exchange, threat, and integrative relationships as the three primary coordinating mechanisms in society.

In the realm of decentralized environmental governance, the hybrid forms of environmental governance depicted in Figure 6.1 also embody a recognition of the significance of Hayek's (1973) "synoptic delusion," in which no single agent can possess the capabilities, knowledge, and wisdom to solve complex, multifaceted environmental problems. Accordingly, reliance on hybrid partnerships seeks to enhance capacity and deepen knowledge in the hope that this will generate solutions to these complex problems.

The genesis of decentralized hybrid governance institutions is comparatively easily explained, but perhaps a more salient question revolves around the efficacy of environmental partnership models. Although the inclusion of multiple parties in environmental decisionmaking undoubtedly enhances the knowledge base on which decisions can be made and provides greater legitimacy for these decisions, hybrid governance models, like all forms of collective endeavor, are subject to the problems posed by individual and group motivation, such as the familiar prisoner's dilemma (as explained below).

The central feature of hybrid partnerships in any sphere of collective decisionmaking may be found in their underlying "structures of resource dependency" (Rhodes 1988). This arises because the groups and organizations that could potentially belong to them control different amounts and types of resources: authority, legitimacy, money, information, and so on. They could thus benefit from engaging in processes of deliberation, compromise, and negotiation that produce a system of horizontal coordination through which dispersed resources can be mobilized and pooled so that "collective (or parallel) action can be orchestrated towards the solution of a common policy" (Kenis and Schneider 1991, 36).

Two major problems would appear to stand in the way of the emergence of this kind of collaboration. The first is the prisoner's, or bargaining, dilemma, which arises in situations where defection from cooperation is more rewarding for opportunistically rational actors than compliance, because of the risk of being cheated (Scharpf 1991). Some actors may withhold the resources they have agreed to contribute to hybrid partnerships and attempt to "free ride" on the contributions other parties make to the advancement of common goals. For instance, the collapse of the Copenhagen climate change conference in 2009 represents a quintessential example of this kind of problem. Second, what Borzel (1998) has termed the "structural dilemma" arises because the actors that engage in partnership decisions are often agents of the groups they claim to represent and are thus subject to pressures from their principals.

Conceptual difficulties facing all collective hybrid partnership attempts at decentralized environmental decisionmaking provide valuable clues on how well these partnerships may work in practice, but an even more important source of insight is real-world experience. A voluminous literature documents literally thousands of case studies (see, e.g., Conley and Moote 2003 for a synoptic review), drawing on the exponential growth of collaborative environmental decisionmaking over the past two decades. Although this literature offers considerable support to efforts at collaborative decisionmaking (see, e.g., Dukes and Firehock 2001), critics abound (see, e.g., Kenney 2000), and they have forced a reevaluation of collaborative environmental decisionmaking. As yet, however, no definitive study has been undertaken, and existing material unfortunately does not use reliable and consistent yardsticks of evaluation (Leach 2000). In a nutshell, doubts are growing, and the jury is still out.

CONCLUSIONS

While we await more definitive empirical evidence on the efficacy of decentralized environmental decisionmaking based hybrid partnership models, where does this leave participants in the Australian debate on water policy, as it is commonly agreed that water problems, including those involved in the recreational and tourism use of water, are pressing? First, it should be stressed that with present knowledge, no great faith should be placed in hybrid partnership models as a panacea to resolve thorny environmental dilemmas that involve "winners" and "losers," because in these kinds of highly charged policy debates, ultimately political decisions are required that rest on judgments on the political strengths of the interest groups at play. However, in stark contrast to command-and-control hierarchies, decentralized hybrid partnership models have a distinct comparative advantage in generating local information on local aspects of broader problems. Accordingly, collaborative hybrid partnerships should be employed, at the very least, to gather and assess decentralized information.

Second, pioneering work by Ostrom (1990) on common-pool resources, such as fisheries, forests, and grazing lands, still has significant practical value in the Australian water milieu. In essence, Ostrom stressed the multifaceted nature of human ecosystem interaction and argued against any single "silver bullet" solution for ecological resource use problems. However, she did offer workable principles for the design of decentralized solutions to these problems, which contemporary Australian water policymakers should bear in mind. Thus, an institutional structure should provide clearly defined boundaries of who can participate in decisionmaking and embody effective exclusion of external parties with insufficient *locus standi*. In addition, any rules regarding the use of common water resources should be adapted to local conditions. Moreover, collective choice arrangements must allow potential users of water to participate in the decisionmaking process. This should be followed up with effective monitoring by agents who are accountable to water users, and water users who violate established water use rules must face sanctions. To minimize the costs of conflict associated with disputes, Ostrom proposed the employment of local mechanisms for conflict resolution that are cheap and easy to access. Finally, to ensure the vibrancy of decentralized decisionmaking entities, the self-determination of the local community must be recognized by higher-level governing authorities.

REFERENCES

Ahmad, E., and G. Brosio. 2009. *Does Decentralization Enhance Service Delivery and Poverty Reduction?*. Cheltenham, UK, Edward Elgar Publishers.

Armitage, D., F. Berkes, and N. Doubleday. 2007. Introduction: Moving beyond co-management, in *Adaptive Co-Management*, edited by D. Armitage and F. Berkes, N. Doubleday. Vancouver, UCB Press, 1–15.

Borzel, T.J. 1998. Organizing Babylon: On the different conceptions of policy networks. *Public Administration* 76:253–273.

Boulding, K.E. 1978. *Ecodynamics*. New York, Sage.

Bradrach, J., and R. Eccles. 1991. Price, authority and trust: From ideal types to plural forms, in *Markets, Hierarchies and Networks: The Co-ordination of Social Life*, edited by G. Thompson and J. Frances, R. Levacic, J. Mitchell. London, Sage, 365–381.

Breton, A. 1995. *Competitive Government*. Cambridge, UK, Cambridge University Press.

Cestero, B. 1999. *Beyond the Hundredth Meeting: A Field Guide to Collaborative Conservation on the West's Public Lands*. Tucson, Sonoran Institute.

Cleary, C.R., and D. Phillippi. 1993. *Coordinated Resource Management*. Denver, Society for Range Management.

Coase, R.H. 1937. The nature of the firm. *Economica* 3: 386–405.

———. 1960. The problem of social cost. *Journal of Law and Economics* 3 (1): 1–44.

Colfer, C.J. 2005. *The Equitable Forest: Diversity, Community and Resource Management*. Washington, DC, Resources for the Future.

Conley, A., and M.A. Moote. 2003. Evaluating collaborative natural resource management. *Society and Natural Resources* 16 (5): 371–386.

Daniels, S.E., and G.B. Walker. 2000. *Working through Environmental Policy Conflicts: The Collaborative Learning Approach*. New York, Praeger.

Dollery, B.E., and J.L. Wallis. 1997. Market failure, government failure, leadership and public policy. *Journal of Interdisciplinary Economics* 8 (2): 113–126.

Downs, A. 1957. *An Economic Theory of Democracy*. Harper and Row, New York.

Dukes, E.F., and K. Firehock. 2001. *Collaboration: A Guide for Environmental Advocates*. Charlottesville, University of Virginia.

EPA (U.S. Environmental Protection Agency). 1997. *Community-Based Environmental Protection: A Resource Book for Protecting Ecosystems and Communities*. Washington, DC, U.S. Environmental Protection Agency.

Gray, G.J., M.J. Enzer, and J. Kusel. 2001. *Understanding Community-Based Ecosystem Management in the United States*. Haworth Press, New York.

Hayek, F.A. 1973. *Law, Legislation and Liberty*. Chicago, University of Chicago Press.

Hutchcroft, P.D. 2001. Centralization and decentralization in administration and politics: Assessing territorial dimensions of authority and power. *Governance* 14: 23–53.

Innes, J.E. 1999. Evaluating consensus building, in *The Consensus Building Handbook*, edited by L. Susskind and S. McKearnan, J. Thomas-Larmer. Thousand Oaks, CA, Sage, 631–675.

Kenis, P., and V. Schneider. 1991. Policy networks and policy analysis: Scrutinizing a new analytical toolbox, in *Policy Network: Empirical Evidence and Theoretical Considerations*, edited by B. Marin and R. Mayntz. Frankfurt, Campus Verlag, 293–412.

Kenney, D.S. 2000. *Arguing about Consensus*. Boulder, Natural Resources Law Center, University of Colorado.

Kenney, D.S., and W.B. Lord. 1999. *Analysis of Institutional Innovation in the Natural Resources and Environmental Realm*. Boulder, Natural Resources Law Center, University of Colorado.

Leach, W.D. 2000. *Evaluating Watershed Partnerships in California: Theoretical and Methodological Perspectives*, PhD diss.,. Department of Ecology, University of California–Los Angeles.

Le Grand, J. 1991. The theory of government failure. *British Journal of Political Sciences* 21 (4): 739–757.

Lemos, M.C., and A. Agrawal. 2006. Environmental governance. *Annual Review of Environmental Resources* 31: 297–325.

Lipsey, R.G., and K. Lancaster. 1956. The general theory of the second best. *Review of Economic Studies* 24 (1): 11–32.

Margerum, R.D. 1999. Integrated environmental management: The foundations for successful practice. *Environmental Management* 24 (2): 151–166.

Mayntz, R. 1993. Modernization and the logic of interorganizational networks, in *Societal Change between Market and Organization*, edited by J. Child and M. Crozier, R. Mayntz. Aldershot, UK, Ashgate Publishing, 3–18.

Moote, A., A. Conley, K. Firehock, and F. Dukes. 2000. *Assessing Research Needs: A Summary of a Workshop on Community-Based Collaboratives*. Tucson, University of Arizona.

NRLC (Natural Resources Law Center). 1996. *The Watershed Source Book*. Boulder, University of Colorado, Natural Resources Law Center.

Oates, W. 1972. *Fiscal Federalism*. New York, Harcourt Brace Jovanovich.

———. 1999. An essay on fiscal federalism. *Journal of Economic Literature* 37 (3): 1120–1149.

O'Dowd, M.C. 1978. The problem of government failure in mixed economies. *South African Journal of Economics* 46 (3): 360–370.

Ostrom, E. 1990. *Governing the Commons: The Evolution of Institutions for Collective Action*. Cambridge, UK, Cambridge University Press.

Ouchi, W. 1991. Markets, bureaucracies and clans, in *Markets, Hierarchies and Networks: The Co-ordination of Social Life*, edited by G. Thompson, J. Frances, R. Levacic, and J. Mitchell. London, Sage, 131–142.

Peltzman, S. 1976. Towards a more general theory of regulation. *Journal of Law and Economics* 19 (2): 211–240.

Rhodes, R. 1988. *Beyond Westminster and Whitehall*. London, Unwin Hyman.

Scharpf, F. 1991. Political institutions, decision styles and policy choices, in *Political Choice: Institutions, Rules, and the Limits of Rationality*, edited by R. Czada and A. Windhoff-Heritier. Boulder, CO, Westview Press, 241–255.

Segerson, K., and T.J. Miceli. 1998. Voluntary environmental agreements: Good or bad news for environmental protection? *Journal of Environmental Economics and Management* 36 (1): 109–130.

Stigler, G.C. 1971. The economic theory of regulation. *Bell Journal of Economics* 2 (1): 137–146.

———. 1975. *Citizen and State: Essays on Regulation*. Chicago, University of Chicago Press.

Streek, W., and P. Schmitter. 1985. *Private Interest Government*. London, Sage.

Tews, K., P.O. Busch, and H. Jorgens. 2003. The diffusion of new environmental policy instruments. *European Journal of Political Research* 42 (4): 569–600.

Thompson, G., J. Frances, R. Levacic, and J. Mitchell (Eds.). 1991. *Markets, Hierarchies and Networks: The Co-ordination of Social Life*. London, Sage.

Vining, A.R., and D.L. Weimer. 1991. Government supply and production failure: A framework based on contestability. *Journal of Public Policy* 10 (1): 1–22.

Wallis, J., and B.E. Dollery. 1999. *Government Failure, Leadership and Public Policy*. London, Palgrave.

Weber, E. 2000. A new vanguard for the environment: Grass-roots ecosystem management as a new environmental movement. *Society and Natural Resources* 13 (3): 237–259.

Weisbrod, B., 1978. Problems of enhancing the public interest: Toward a model of government failures, in *Public Interest Law*, edited by B. Weisbrod, J. Handler and N. Komesar Berkeley, University of California Press, 30–41.

Western, D., and Wright, R.M. (Eds.), 1994. *Natural Connections: Perspectives in Community-Based Conservation*, Washington, DC, Island Press.

Williams, E.M., and P.V. Ellefson. 1997. Going into partnership to manage a landscape. *Journal of Forestry* 95 (5): 29–33.

Williamson, O.E. 1985. *The Economic Institutions of Capitalism*. New York, Free Press.

Wolf, C. 1979a. A theory of non-market failure: Framework for implementation analysis. *Journal of Law and Economics* 22 (1): 107–139.

———. 1979b. A theory of non-market failures. *Public Interest* 55 (2): 114–133.

———. 1983. 'Non-market failure' revisited: The anatomy and physiology of government deficiencies, in *Anatomy of Government Deficiencies*, edited by H. Hanusch. New York, Springer-Verlag, 138–151.

———. 1987. Market and non-market failures: Comparison and assessment. *Journal of Public Policy* 6 (1): 43–70.

———. 1989. *Markets or Governments*. New York, MIT Press.

Zerner, C. 2000. *People, Plants and Justice*. New York, Columbia University Press.

CHAPTER 7

Collaborating and Coordinating Disparate Interests: Lessons from Water Trusts

Sue O'Keefe and Brian Dollery

C hapter 6 provided a survey of the literature on public policymaking and introduced models of collaboration in water policy, administration, and usage. Noting the many weaknesses of a top-down approach to environmental problems, it introduced a number of models of collaboration, including that of collective hybrid partnerships. In essence, that chapter set the scene for consideration of the specific form of market-based collaboration in this chapter.

Collaborative models are likely to be of particular interest to the tourism and recreation sector, which, at best, has had only a modest influence on water policy in Australia, largely because of the difficulties associated with precisely defining the sector and with the disparate interests that it represents. Moreover, as noted earlier in this book, understanding of the impacts of modifications to hydrological systems is incomplete, and both conflicts and complementarities exist among consumptive and nonconsumptive users. Accordingly, policymakers have focused, perhaps understandably, on those interests most directly affected by reduced supplies, where market values are revealed, knowledge is most complete, and the interest is most vociferously represented at the policy table. As noted in Chapter 1, this has largely precluded the tourism and recreation sector.

The context of water policy in Australia is heavily influenced by the spatial and temporal variability of the resource, which has led to considerable efforts to make water supplies more reliable or secure. As noted in earlier chapters, the tension between the requirements of agriculture and the environment has resulted in costly modifications to natural flows, which have posed serious threats to ecological systems. Nonetheless, increased knowledge and public concern about the environmental impacts of overallocation has resulted in increased recognition of the benefits of instream flows, and governments are now actively involved in purchasing volumetric rights in the Murray-Darling Basin (MDB) to achieve preferred environmental outcomes and return extractions to more acceptable limits. At present, as part of the government's A$12.9 billion Water for the Future plan, A$3.1 billion over 10 years has been committed to buying back water entitlements.[1] The likelihood is strong that a very large volume of water will be designated for the environment in coming years. In fact, in the collection of papers assembled by the Australian Farm Institute (Bennett et al., 2010) it is predicted that the Commonwealth Environmental Water Holder will become the single largest owner of water in the MDB, controlling up to 20% of extractive water rights. About 797 gigaliters of water entitlements have been acquired as of February 2010 under the federal government's Restoring the Balance program (Productivity Commission, 2010).

This situation is not dissimilar to that facing the western United States, where, on the back of significant reforms establishing and developing water markets and growing public support for environmental values, attention has increasingly turned to more market-based strategies to achieve environmental outcomes. This shift has been termed "market environmentalism" (Anderson and Leal, 1991) or "private conservation" (Del Alessi, 1999), and it has seen the development of alliances between environmental and recreation interests in the form of water trusts that actively participate in water markets to secure increased instream flows. In the western United States, private water trusts have resulted in enhanced environmental and recreational outcomes and have also revealed the market value for nonconsumptive water use.

In this chapter, we draw on the experience of water trusts in the United States and similar bodies in the field of land conservation in Australia to examine potential collaborative models for the tourism and recreation sector. The chapter begins with a brief outline of the American and Australian contexts. This is followed by an examination of the concept of market environmentalism, focusing on the operation of water trusts in the United States and land and environmental trusts in Australia.

BACKGROUND AND CONTEXT: AUSTRALIA AND THE UNITED STATES

As noted in Chapter 5, in the context of extreme variability of water supplies, the definition, monitoring, and enforcement of water rights become crucially important. Water rights facilitate the efficient use of water and its ongoing transfer to more highly valued uses. Well-defined property rights also assist in achieving a balance among economic, social, and environmental interests (Libecap et al., 2009). The establishment and development of water markets in Australia and the United States offer evidence of the potential for water rights and markets to "provide information on current consumption patterns and alternative values, incentives for adjustments in use, and smoother reallocation across competing demands" (Libecap et al., 2009, 1). In both countries, we have witnessed the role of rights and markets in enhancing protection for instream flows.[2]

Although different rights systems prevail across the two countries, Libecap and colleagues note several similarities in background and context. These include climate variability and the accompanying need for reservoirs to ensure reliability of supply; the need for cross-border water management; a history of the majority allocation going to irrigated agriculture; increasing competition among different uses, including agricultural, environmental, recreational, and urban residential; and the potential for trade to enhance allocation across these uses. Moreover, in each country, the environmental movement and the broader move to a service economy has redirected conservation goals toward recreation and ecosystem protection and has led to recognition of the importance of instream flows (King, 2004; Libecap et al., 2009).

In Australia and the United States, ownership of the water resource is vested in the state, but other users have historically been granted rights over its use, albeit subject to certain state-imposed conditions. In the MDB, a system of statutory rights prevails, whereas in the western United States, appropriative rights govern water markets. The following sections provide brief explanations of these rights.

Water Rights in the Murray-Darling Basin

Historically, Australian states have primarily seen their water rights as a means to develop irrigated agriculture with the free allocation of statutory water rights, typically 1 acre-foot, along with state-sponsored irrigation infrastructure. By the 1980s, however, the serious overallocation of rights led to increasing pressure to decouple land rights from water rights to allow for trade. The result was the establishment of water markets in South Australia in 1982, New South Wales and Queensland in 1989, and Victoria in 1991 (MDBC 1998). The Coalition of Australian Governments in 1994 led to the separation of all land from water rights, and in 1994, the Water Reform framework recognized the need to protect the aquatic environment and called for each state to allocate water to this end (Postel and Richter, 2003). Despite this, all states developed rules to prevent or limit interstate or interbasin transfers. The Water Act 2007 establishes the Commonwealth Environmental Water Holder as the agency responsible for managing the water entitlements that the commonwealth is currently acquiring through the Restoring the Balance program in the MDB (water entitlement purchasing) and Sustainable Rural Water Use and Infrastructure program.

Although a uniform framework exists for achieving reform, and the Water Act foreshadows greater control by the commonwealth, each state has retained jurisdictional nuances that substantially complicate the policy and market environment. Calls for uniformity are relatively common in the popular media, although the usefulness of these measures is often overstated. For example, there is no compelling evidence to suggest that the politics of water is any less problematic at a national level than at a state or regional level. Furthermore, as noted in Chapter 6, disenchantment with a top-down approach has been well documented.

A form of uniformity of statutory rights exists in the principles that underpin the National Water Initiative. More specifically, all jurisdictions have agreed that any consumptive use of water requires a water access entitlement that is to be described in legislation as a perpetual share of the consumptive pool of a water resource.[3] However, the priority of rights can vary across and between jurisdictions.

For example, New South Wales has high-security and relatively low-security right holders, whereas in South Australia most rights are relatively uniform in priority.

Water Rights in the United States

Appropriative rights involve the separation of water rights from landownership, and in the western United States, they are based on the "first in time, first in right" rule. The result is a "ladder of rights" (Kwasniak, 2006), with the most senior rights for beneficial use at the top. A right holder has the right to divert a certain amount of water for beneficial use, but the amount of diversion depends on streamflow and reservoir size. Those with the earliest water claims, often irrigators, have highest priority, while those with subsequent claims, such as urban users, have lower priority or junior claims. In this context, trade allows those with junior rights but high-value water uses (urban users are a typical example of this) to lease or purchase water from those with lower-valued uses but higher priority rights. The resulting allocation is to their mutual advantage.

The requirement of beneficial use has historically been seen as synonymous with consumptive use as measured by physical diversions. The unfortunate result of the application of the beneficial use test has been the creation of incentives for users to use water in low-value ways simply to conserve their rights (King, 2004). Water that has been conserved or salvaged therefore would not count as beneficial use, and irrigators are perversely encouraged to "use it or lose it."

Until recently, environmental uses or instream flows were not recognized as beneficial use, to the obvious detriment of fish stocks, recreation and other environmental values. King (2004) has identified several necessary preconditions for the incorporation of instream values: the presence of water markets, advanced scientific knowledge, institutional effectiveness, and the political environment. Legislation to protect instream flows recognizes the public benefit of environmental amenity, and this has paved the way for private and semipublic organizations, such as the Oregon Water Trust (discussed below), to enter the market to maintain streamflows and protect recreational and environmental interests.

The similarities between the two countries raise an important possibility. In each nation, water for environmental purposes has purportedly been purchased by the government, but the roles for private organizations and agencies differ. The United States has embraced a form of market environmentalism that removes some of the difficulties associated with government action on buyback and has resulted in an enhancement of environmental and recreational interests. Movement on this front has been modest in Australia; however, in considering the applicability of private agent activity to secure environmental services, the Productivity Commission (2009) recently highlighted the potential of private trusts based on both altruistic motives and the recreational value of water. More recently, Bennett et al. (2010) have also highlighted the potential for devolved decisionmaking on the environmental front, along with the potential role for trusts to play in allocating environmental water in Australia. The following section briefly outlines the concept of market environmentalism to provide the background for assessing its relevance in Australia.

PRIVATE CONSERVATION/MARKET ENVIRONMENTALISM

Chapter 6 considered three generic types of environmental management tools and focused in particular on market incentives in the context of decentralized environmental management models. The economic advantages of markets are well documented, and market-based instruments increasingly are being developed and used in conservation and environmental protection efforts. Market environmentalism rests on the assumption that markets are essential to the efficient allocation of natural resources, and market advocates question the government's ability to respond effectively and efficiently to market failures such as information costs and third-party effects (Anderson and Snyder, 1997). Proponents hold that private organizations, like land or water trusts, have an important role to play and can bring together environmental, philanthropic, and recreational interests (Binning and Feilman, 2000; Del Alessi, 1999). According to Del Alessi (1999, 2), the prevalent government-based approach is built on political institutions that "create more conflicts than they solve" because, in contrast to the marketplace, where voluntary trade benefits each party, politics is a zero-sum game where the gains of one group are necessarily made at the expense of the other. It is argued that private ownership changes the incentives people face and results in more creative solutions.

In Australia, this approach is most developed in landscape conservation, which involves the nongovernment sector purchasing land. In the United States, however, private water market transactions are increasingly being used to secure improved outcomes for the environment and recreational and tourism interests. Private water rights systems use market transactions to secure environmental flows in situations in which highly valued environmental or recreational services are threatened (Katz, 2006), and the result is revealed market values for instream flows in excess of those for agriculture (Colby, 1990).

Water Trusts in the United States

Until relatively recently, it was thought that any water left instream was water wasted, but the recognition of instream use as beneficial paved the way for the development of water trusts in the western United States. "Water trusts are private, non profit organisations that acquire water rights in order to enhance instream flow for conservation purposes" and play an important part in the restoration, maintenance, and enhancement of aquatic and riparian ecological integrity (King, 2004, 211). In states such as Washington and Oregon, private organizations may acquire instream water rights but are prohibited from holding them themselves. These rights must be transferred to and held by the state (King, 2004). In essence, this system results in the acquisition and transfer of rights from agricultural to instream use, to the benefit of both the environment and recreational users. Under the umbrella of incentive-based conservation, this involves harnessing a market-based approach to address environmental concerns. Water trusts are seen by King as a valuable tool to promote conservation such as enhancement of fisheries, water quality, habitat, and recreation.

The initial preference of water trusts was to purchase rights outright to achieve a permanent transfer (Landry, 1998), but the transfer of appropriative rights is subject to numerous and at times prohibitive rules (Productivity Commission, 2009). For example, the California constitution makes it illegal to acquire water rights to which someone is entitled, although the impact of this has been reduced by an expansion of the public trust doctrine to include not only environmental values, but also recreational and aesthetic ones (Productivity Commission, 2009). Transfer of rights to the state can also involve protracted legal processes, making leasing a more feasible and practical option in many cases. The result has been that in practice, the majority of transactions completed have been leases, which result in the temporary reallocation of water rights. Loomis et al. (2003) recognize that leases offer several advantages: they are temporary and offer flexibility for both parties; rights holders can become comfortable with the idea of instream flow marketing and have a chance to see how the lease affects their water needs; leasing organizations can assess the effectiveness of the volume leased in protecting instream flows; and, importantly, observers have the opportunity to determine whether the transfer of water from agriculture does in fact have a detrimental impact on agricultural communities, which has been one of the greatest fears expressed by opponents of trade. Moreover, a variety of leasing options are available to accommodate the needs of buyers and sellers. These include standard annual or multiyear leases, as well as dry-year leases, which include prior arrangements for access to the water right in a dry year (similar to options contracts), and split-season leases, which allow a portion of the water right to be used in irrigation early in the year, with the remainder of the right for instream use in the summer (Landry, 1998). Leases also enable the rapid transfer of water, in contrast to permanent transfers, which in Oregon can take a number of years (Productivity Commission, 2009). Accordingly, water trusts make limited use of tax concessions, as only a permanent transfer of a water right qualifies as a tax deduction (King, 2004).

Reliance on voluntary transfers of water can complement state instream flow programs but also bring substantial benefits compared with the "command and control" of regulation or compulsory acquisition. Trusts can potentially overcome inadequate funding, ineffective enforcement, and the procurement of junior rights, and they can circumvent the slow and expensive bureaucratic process and bring agility to projects though speed, flexibility, and creativity (King, 2004). They are also able to raise funds from those more comfortable working with the nonprofit sector than with the government sector; provide resources in the form of people, institutions, infrastructure, and funding; and bring social risk capital (O'Neill, 1989).

Water trusts bring potential solutions to the failure of public agencies to protect instream flows, and institutionally, they divert much of the opposition and hostility away from the state, especially from agricultural interests. King (2004) also identifies a symbiotic relationship between water trusts and the state, with trusts providing much of the site-specific knowledge and groundwork, while the state provides the legal, institutional, and physical infrastructure to ensure the transfer of the water right.

The incentives for purchasing water rights to protect instream flows include philanthropic, legal, and monetary ones. Tarlock and Nagel (1989) note that both self-interest and public acceptance are important elements of this type of activity, while King (2004) identifies altruism along with the protection from forfeiture due to lapsed usage as salient incentives to participate in water markets. Selling or leasing water in cases where crops are unprofitable or weather conditions adverse also makes sound economic sense, as does monetary compensation or, as occurs in some cases, sponsored improvements in irrigation.

The Freshwater Trust is one of the better-known water trusts in the United States. It was formed in 2009 through the merger of the Oregon Water Trust (OWT) and Oregon Trout (OT). The OWT was the first nonprofit private water trust in the United States, formed in 1993. Its aim was to restore surface-water flows using cooperative market-based solutions in areas targeted for maximum ecological benefit. OT had been formed earlier, in 1983, by a group of fly-fishing conservationists to protect and restore native fish and their ecosystems. The Freshwater Trust takes an "integrated, innovative approach to restoring freshwater ecosystems—from restoring a river's architecture to working with landowners to keep more water in streams to educating children on the importance of freshwater conservation" (Freshwater Trust, 2010). It achieves its goals through the purchase, lease, or donation of water rights, usually from irrigators. The water gained is then applied to restore site-specific amenity to the benefit of the environment, recreational interests, and communities alike. Notably, this trust uses market-based incentives, but it also brings together scientific knowledge and management expertise and sees its charter as including an educative role. It is widely cited as an example of the potential for market-based solutions to problems of water allocation, or incentive-based conservation (King, 2004; Kwasniak, 2006).

Oregon, in particular, relies on individuals and groups to buy environmental water rights, which are then, by law, donated to the state government (Productivity Commission, 2009). Thus, the Oregon government does not buy or lease water directly, but instead relies on donations from conservation groups such as the Freshwater Trust and the Deschutes River Conservancy, although these groups are often subsidized by the state and federal governments. In 2006, the Freshwater Trust had about 390 megaliters per day of instream flows acquired through donations or payments to landholders, of which 245 megaliters were leased from landholders or irrigation districts, with the remainder purchased (Productivity Commission, 2009). Most purchases are from individual land-holders, but the trust also leases from irrigation districts.

Rather than directly approaching landholders, more success has been achieved by working with local conservation groups, who introduced the Freshwater Trust to those interested in selling water (Productivity Commission, 2009). Most acquisitions are negotiated and offer considerable flexibility depending on the landholder's circumstances, the farmer's environmental or recreational objectives, and the specific streamflow demands evident. This has resulted in the crafting of mechanisms such as split-season leases, where an agreement is reached so that the timing of irrigation flows imposes least cost on the environmental asset. The Freshwater Trust has also used "in-kind" payments of hay for water.

Australian Incentive-Based Environmentalism

In her analysis of the foundations of the US water trust movement, King (2004) notes that it was built on the success of the land trust movement. King points out three central attributes of the land trust movement seen to underpin its successful application to environmental challenges: first, it draws on the theory of incentive-based conservation; second, it operates in perpetuity; and third, it operates as a public-private partnership. These three elements are also present in Australian land trusts that have evolved in attempts to protect biodiversity and secure ecological outcomes. The following section briefly describes some of the key players in market-based conservation in Australia.

Australian Wildlife Conservancy (AWC) is the largest nongovernment owner of conservation land in Australia (Totaro, 2009), with ownership of 21 sanctuaries covering 2.5 million hectares in places such as north Queensland, the Kimberley, western New South Wales, Northern Territory, and the forests of southwestern Australia (AWC 2009). It is focused not only on the purchase of land, but also on public education and on-ground support that is backed by a substantial research focus. AWC aims to ensure that its sanctuaries act as "catalysts" for broader landscape-scale conservation efforts. Accordingly, the conservancy works closely with its neighbors to also promote conservation beyond the borders of each AWC sanctuary.

Bush Heritage was originally set up by Bob Brown to protect old-growth forest in Tasmania and has developed to buy and manage land of high conservation value throughout Australia. It also works through building partnerships with other landholders and derives 55% of its funding from individual donations, with other, relatively small contributions from corporations, bequests, and grants (Bush Heritage, 2010).

These examples indicate a growing role for land trusts in Australia, but the same cannot be said of water trusts. Several reasons may feasibly account for this, and many of these reasons are expanded on throughout this book. Explanations may reside in the tangibility and maturity of the land market in comparison to the relatively new water market; lack of knowledge about the impacts of flow regimes and timing; specification of water rights in terms of volume only, which does not necessarily allow for simple calculations of the extent of overlap between environmental and recreational benefits; and the relatively recent recognition of nonconsumptive values of water for environmental and recreational interests. Nonetheless, the success of water trusts in the United States along with that of land trusts in Australia may give some hope for the future establishment and development of private organizations in Australian water markets. For example, some environmental organizations in Australia, such as Healthy Rivers, accept donations of water rights, and to date they have played a modest role in the overall market for environmental flows. The Productivity Commission suggests possible explanations for the difference in approach between Australia and the United States. First, it is possible that members of US trusts are able to capture most of the benefits of improved environmental outcomes, such as through recreational fishing; second, it is also possible that the provision of environmental services is seen as traditionally the domain of government. The Productivity Commission further notes the potential for government activity in the market to crowd out private participation (2009).

Notwithstanding these impediments, observers have recently noted the potential advantages of trusts in an Australian context. For instance, Young (2010) argues that the environmental water reserve should be specified as entitlements that would then be devolved and held by local environmental trusts or similar agencies. The role of the central water holder would be limited to holding only those entitlements that could not be sensibly managed by local entities. Young notes that the underlying rationale for these arrangements is the subsidiarity principle, which is discussed in Chapter 6 of this book. Under Young's regime, local trusts (such as irrigators) would know how much water they had at their disposal each year and could allocate that water to achieve their objectives. This could feasibly include buying and selling allocations. While this approach is still in embryonic form, several vehicles for private activity to secure environmental and recreational flows already exist in Australia, with Healthy Rivers Australia's Water Bank and the Murray Wetlands Working Groups providing some useful insights.

The aim of Healthy Rivers Australia is to build the country's largest environmental water bank (HRA 2010b), with contributions coming from any source and including donations of water and money to purchase water. It is a not-for-profit independent membership-based organization that works with communities to restore river health. The Water Bank holds environmental water in trust for projects that improve water quality or quantity and the environment for native flora and fauna. It is independent of government, and tax-deductible financial donations are managed by a committee of community members. In 2008, despite its aims and potential, only 4 megaliters of water was delivered for environmental outcomes (HRA 2007–2008). However, the organization has recently achieved what is seen as a breakthrough in achieving a tax concession status. Although this type of incentive has existed for land conservation groups, this has previously not been the case for water. In June 2010, Healthy Rivers Australia helped a New South Wales irrigation license holder receive a tax concession of A$16,900 for a temporary donation of 48.4 megaliters of water (HRA 2010a) targeted for wetland restoration and fish reintroduction in the Murray. Moreover, Healthy Rivers Australia received 59.75 megaliters in 2008–2009 (HRA 2010a), with particularly busy activity in the last two weeks of the season, as irrigators realized that they were unlikely to be able to find an alternative use for small amounts of water.

New South Wales Murray Wetlands Working Group (MWWG) has the aims of rehabilitating degraded wetlands and improving wetland management for Murray and Lower Darling catchments (Nias, 2010). The MWWG was formed in 1992 as an initiative of the Murray and Lower Murray-Darling Catchment Management Committees. Its members include private irrigators and landholders, shire councils, Catchment Management Authorities, Freshwater Research Centres, state and federal government departments, and independent ecologists. Its focus is on a targeted approach to conservation to maximize benefits. The MWWG is an incorporated group that has its own constitution and develops its own charter of activities. However, more recently, a new group, Murray Darling Wetlands Ltd., has been formed as a private company, similar in nature to US water trusts or Australian irrigation trusts. This initiative has allowed for donations to be tax-deductible. In the past, groups such as these have been discouraged from entering the water market, largely because of the restrictive rules around the transfer of water rights, but there is some hope that these rules will be relaxed in future. The MWWG has received interest from the tourism and recreation sector, with members clearly discerning some convergence of interests similar to that experienced in the United States. A further market is being explored that takes in the corporate sector and taps into the corporate environmental sustainability movement in much the same manner as many organizations currently support tree plantings. Changing to a private entity would allow tax-deductible status for these corporate partners. The trust model offers increased flexibility compared with government programs and provides expanded potential to build relationships within communities.

A recent initiative of the MWWG has been the establishment of the Water Trust Alliance, along with five other nongovernmental organizations, to strengthen the role of community in securing environmental outcomes for Australia's rivers. The Water Trust Alliance will act as a national-level peak body for local water trusts, coordinating the latest information and best practices in water trust models, water entitlements, trust acquisition, and management. By developing strategic connections with governments, landowners, businesses, and community groups across a wide region, the alliance will enable water trusts to deliver effective environmental water management (ACF 2010).

CONCLUSIONS

The extension of water market activity to include the interests of tourism and recreation alongside environmental interests appears to offer some scope for improved instream flows. This approach to conservation is well established in the United States, and in an Australian context, it exists primarily in the form of land or nature conservation trusts. However, the precise operation and design of water market programs are intricately connected to the specific institutional and legal context. Hence, just as differences exist in the operation of these organizations across the western United States, there are likely to be particular impediments and enabling factors within Australian jurisdictions. These characteristics lessen the likelihood of a simple transfer of institutional arrangements across jurisdictions.

In Australia, the problems of overallocation and associated environmental degradation have been recently addressed through government buyback programs. However, these have attracted widespread criticism from agricultural interests and more recently the Productivity Commission (2009). Despite government assurances that it will purchase water for the environment only from willing sellers, the agricultural community increasingly fears that a more targeted approach will be taken, with some irrigation communities suffering inordinately. From another perspective, environmentalists such as the Australian Conservation

Federation prefer a targeted approach to ensure the most impact from any purchases of environmental water. In contrast, a mutually satisfying market-based trade leaves both parties better off and avoids the "divisive, time consuming rancor" that accompanies state intervention in water reallocation (King, 2004). In this context, there may be scope for expanded efforts of private nonprofit entities to improve ecological outcomes. These organizations have several advantages over government activity, not the least being that they are more likely to receive support from an irrigation community mistrustful of government meddling. Clearly, the development of water trusts in Australia would give scope for the recreation and tourism industries to influence water allocation through their direct action in the water market as opposed to indirect political lobbying at an already crowded policy table.

From the perspective of the recreation industry, participation in the water market has the advantage of revealing the market value for recreation, in contrast to the current situation in which recreational values are not observable in the market. To date, information about recreational and environmental values has been obtained through the employment of a number of techniques for the nonmarket valuation of the benefits. In the United States, where market activity has occurred through water trusts, the economic value of water for recreation has been shown in some contexts to be at least four times that of agriculture (Ward, 1989). Further potential for benefits to recreation are also noted by Ward, who maintains that even without recreation directly buying water rights, irrigation districts could draw down first from those reservoirs that contribute least to marginal recreational benefit. The efficient timing of irrigation demands would minimize recreational benefits forgone to the regional economy.

Despite the potential advantages from the development of water trusts in Australia, several concerns remain unresolved, and many of these are addressed in detail in other chapters in this book. First, knowledge is incomplete about the extent of complementarity between the water needs of the environment and the various types of recreational values. Second, institutional and legislative arrangements require further investigation to ascertain the potential for water trust activity in Australia. Third, public knowledge and awareness of the potential of water markets are lacking. Fourth, specification of rights in volumetric terms limits the potential impact of water right purchases for both environmental and recreational purposes. Finally, the current bans on interbasin transfers severely limit the potential for improved outcomes for both environmental and recreational uses. An expanded research effort that addresses these concerns and a close examination of alternative institutional structures appear warranted in an effort to more fully understand the potential for increased involvement of the tourism and leisure sector.

NOTES

1. A\$1 = US\$0.9978 as of January 2011.

2. It is also worth noting that some perverse outcomes have also been witnessed. In some instances, increased extraction resulted from the (predictable) activation of "sleeper and dozer

rights," which are water rights that have not been in use or have only occasionally been in use. Trade in these rights has led to an increase in extractions as formerly inactive rights were sold and activated.

3. National Water Initiative, para. 28.

REFERENCES

ACF (Australian Conservation Foundation). 2010. *Water Trust Alliance to Strengthen Role of Community*, accessed October 5, 2010, from www.acfonline.org.au.

Anderson, T.L., and D.R. Leal. 1991. *Free Market Environmentalism*. Pacific Research Institute for Public Policy. Boulder: Westview Press.

Anderson, T., and P. Snyder. 1997. *Water Markets: Priming the Invisible Pump*. Washington DC: Cato Institute.

AWC (Australian Wildlife Conservancy). 2009. *What Does AWC Do?* accessed October 19, 2010, from www.australianwildlife.org.

Bennett, J., R. Kingsford, R. Norris, and M. Young. 2010. *Making Decisions about Environmental Water Allocations*. Surry Hills, New South Wales: Australian Farm Institute.

Binning, C., and P. Feilman. 2000. *Landscape Conservation and the Non-government Sector*. Research Report 7/00, Canberra, Environment Australia.

Bush Heritage. 2010. *About Us*, accessed October 24, 2010, from www.bushheritage.org.au.

Colby, B. 1990. Enhancing instream flow benefits in an era of water marketing. *Water Resources Research* 26 (6):1113–1120.

———. 1999. *Private Conservation Markets, Politics and Voluntary Action*. Hal Clough Lecture for 1999. Edited by M. Del Alessi. Melbourne: Institute of Public Affairs.

Freshwater Trust. 2010. *About Us*. accessed October 24, 2010, from www.thefreshwatertrust.org.

HRA (Healthy Rivers Australia). 2008. *Annual Report 2007–2008*. accessed September 10 from www.healthyrivers.org.au.

———. 2010a. Personal communication with the authors, September 1, 2010.

———. 2010b. *Water Bank*, accessed October 19, 2010, from www.healthyrivers.org.au.

Katz, D. 2006. Going with the flow: Preserving and restoring instream water allocations. In *The World's Water: 2006–2007*. Chicago: Island Press.

King, M. 2004. Getting our feet wet: An introduction to water trusts. *Harvard Environmental Law Review* 28:495–534.

Kwasniak, A. 2006. Quenching instream thirsts: A role for water trusts in the Prairie Provinces. *Journal of Environmental Law and Practice* 16 (3):211–237.

Landry, C. 1998. *Saving Our Streams through Water Markets: A Practical Guide*. Bozeman, MT: Political Economy Research Center.

Libecap, G.D., R.Q. Grafton, C. Landry, and J.R. O'Brien. 2009. *Markets—Water Markets: Australia's Murray-Darling Basin and the US Southwest*. Working Paper No. 15. Prague: International Centre for Economic Research.

Loomis, J.B., K. Quattlebaum, T.C. Brown, and S.J. Alexander. 2003. Expanding institutional arrangements for acquiring water for environmental purposes: Transactions evidence for the western United States. *International Journal of Water Resources Development* 13 (1):21–28.

MDBC (Murray Darling Basin Commission). 1998. *Managing the Water Resources of the Murray-Darling Basin*. Canberra: Murray Darling Basin Commission.

Nias, D. 2010. Personal communication with the authors. August 21.

O'Neill, M. 1989. *The Third America: The Emergence of the Non-profit Sector in the United States*. San Francisco: Jossey-Bass.

Postel, S., and B. Richter. 2003. *Rivers for Life: Managing Water for People and Nature*. Washington, DC: Island Press.

Productivity Commission. 2009. *Market Mechanisms for Recovering Water in the Murray-Darling Basin: Draft Research Report*. Melbourne: Productivity Commission.

————. 2010. *Market Mechanisms for Recovering Water in the Murray-Darling Basin: Final Research Report*. Melbourne: Productivity Commission.

Tarlock, A.D., and D.K. Nagel. 1989. Future issues in instream protection in the West, in *Instream Flow Protection in the West*, edited by L.J. Macdonnell, T.A. Rice, and S.J. Shupe. Boulder: Natural Resources Law Center, University of Colorado School of Law, 137–155.

Totaro, P. 2009. Flannery takes conservation plea to Europe. *The Age* :7 October 9.

Ward, F. 1989. Efficiently managing spatially competing water uses: New evidence from a regional recreational demand model. *Journal of Regional Science* 29 (2):229–246.

Young, M. 2010. Managing environmental water, in *Making Decisions about Environmental Water Allocations*, edited by J. Bennett, R. Kingsford, R. Norris, and M. Young. Surry Hills, New South Wales: Australian Farm Institute, 1–80.

PART III

PRACTICAL CHALLENGES
AND POLICY FORMULATION

CHAPTER 8

The Swan River: Look but Do Not Touch

Fiona Haslam McKenzie

*I*n many parts of the world, within the context of broader economic restructuring, urban river systems have been reinterpreted as resources of value for the purposes of recreation and tourism (Marzano et al. 2009). This is not the case in Perth. Almost every visitor here will, at some time during his or her visit, look at and admire the section of the Swan River in Perth that passes in front of the central business district (CBD), commonly referred to as Perth Water. It is more than likely, however, that this is all he or she will do—look but not touch or really experience all that the Swan River has to offer in terms of its recreational potential, history, or heritage.

Compared with other Australian cities, such as Sydney, Melbourne, and Brisbane, tourism is "underdone" in Perth. Approximately 3.5 million people visit Perth a year, and not surprisingly, international visitors stay the longest and spend the most while they are visiting (Tourism Western Australia 2009). Brisbane, by comparison, has a similar population, but about 5.5 million people visit each year and spend almost twice as much per capita (Tourism Queensland 2009). Melbourne has twice the population and, it can be argued, does not have proximity to superb natural assets such as both Brisbane and Perth have, but tourism is a very important industry for the city and Victoria more broadly. Neither the Yarra River in Melbourne nor the Brisbane River in Brisbane provides the open water vistas or recreational variety potentially available on the Swan River. Nonetheless, both cities aggressively pursue tourism opportunities in and around the city centers and the rivers that flow through them.

In examining the nexus between water and tourism, the Swan River makes a compelling and timely case study. First, the Swan River was named as Western Australia's first official heritage icon in 2002 by Premier Geoff Gallop. Second, more tourist hotels and services are concentrated in the CBD than anywhere else in Perth. Third, Perth Water is one of the largest stretches of water in the entire metropolitan river system. This research was conducted against the backdrop of considerable public debate about the lack of foreshore development at Perth, the characterization of Perth as "Dullsville" in the media, and community concern for the health of the Swan-Canning River system.

This chapter focuses on why, despite its potential, Perth Water remains an underdeveloped and undervalued tourism asset. It is argued in this chapter that development on the Swan River, particularly that related to tourism, is constrained by inertia and NIMBYism, which has broad economic and social implications.[1] In contrast, this chapter also examines the process behind the recent revitalization of the Harlem River in New York City in an effort to derive lessons to assist decisionmakers in using Perth Water more presciently.

This chapter first presents the background and context, including the modern history of Perth Water, and reviews the current tourism opportunities. Issues relating to access to Perth Water are then raised, followed by an examination of public perceptions of Perth as "Dullsville" and a "nanny state." Next the chapter looks at the governance and perceived "ownership" of the river in light of progress, or lack thereof, in developing infrastructure on the river while also rehabilitating and preserving the riparian ecosystems. By comparison, the development of an enhanced tourist experience in Harlem, New York, is then examined. In drawing people to the waterways, Harlem has ignited a sense of local ownership and, consequently, a willingness for public moneys to be continually invested in its natural assets to ensure their health and attractiveness for sustained utilization.

BACKGROUND AND CONTEXT

For more than 175 years, the Swan River has been important for a variety of reasons, first as a vital source of food and transport, both before and after the English colonists arrived in 1829, and later as a precious recreational venue for boat enthusiasts, swimmers, walkers, cyclists, and participants in numerous other water-based activities. The Swan and Canning Rivers, which drain the Avon and Swan coastal catchments, flow through the heart of metropolitan Perth. The Canning River joins the Swan River just below the CBD. The two rivers have been of great importance to the local Aborigines, the Nyoongar people, not only as a source of food, but also for their spiritual significance. The Swan-Canning system and the nearby coastal plain therefore represent a cultural, historical, economic, and recreational focus for Western Australia.

The river, particularly that section around the Perth CBD, is indeed a beautiful natural asset. Proximity to, and a view of, the Swan River adds a considerable premium to the value of property, and important business and tourist accommodations are located in the Perth CBD. Perth Water is generally shallow and varies in width, ranging from a relatively narrow waterway at the northern end near the Causeway and Burswood Casino to a wide body of water in front of the city, narrowing at the foot of Kings Park at the Narrows Bridge before spilling into a larger expanse of water referred to as Melville Water at the southwestern end of the city center (see Figure 8.1). Above Perth Water, overlooking the CBD, is

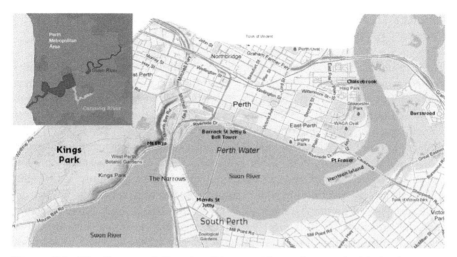

Figure 8.1 *The Swan and Canning Rivers, with Perth water highlighted*

Mount Eliza, the focal point for Kings Park (Seddon 1970), another natural icon to which tourists flock and the single most visited place in Perth by tourists. As the preeminent historian and naturalist George Seddon notes, "No other capital city in Australia has such a constant and universally preferred point of vantage" (Seddon and Ravine 1986 17). Here, with stunning views over the city and the river, are a large expanse of preserved bush, a number of cafés, lookouts, an upscale gift store, and a fine-dining restaurant. It is well serviced by public transport and easily accessible from the city center. It is not, however, easily accessible from Perth Water.

At the eastern end of Perth Water is Heirisson Island, once a series of islands and muddy shallows; these were later dredged to facilitate navigation, and the islands were linked into one (Seddon 1970). Heirisson Island is occasionally referred to as the "Gateway to Perth," but to date, no development has been done on the island, even though several master plans have been submitted. The most recent, a plan to develop the island as a sculpture park with provision for low-impact recreation, was submitted in 2008 (Urbis 2008). Tourists currently have no encouragement to go there; with poor access and limited signage, it is simply a piece of land to be traversed to get to the city center, even though it has a statue of a significant Aboriginal man, Yagan, the hero of resistance to the white colonial invaders. Opposite Heirisson Island on the Perth side is Point Fraser, a short promontory that had been used as a dumping ground for reclamation and dredging work until 2006, when it was rehabilitated to characterize the area prior to colonization, including restoration of the original indigenous dryland and wetland flora of the area (see City of Perth Council 2009). An aquatic ecosystem was re-created to demonstrate how a wetland can be an asset to urban living and raise awareness about local flora and fauna. Point Fraser also has barbecue facilities, a parking area, and a bicycle rental facility. This award-winning work is of potential interest to tourists, but educational tours must be arranged 14 days prior to the proposed tour date, which precludes visitor spontaneity.

RECREATIONAL OPPORTUNITIES FOR TOURISTS ON THE SWAN RIVER

The concentration of hotels and other tourist facilities in the Perth CBD makes Perth Water an important gateway and tourist dispersing point. However, a review of the options available to a tourist who would like to recreate on or by the Swan River at Perth Water shows that the choices are relatively limited. The most notable structure, at the foot of Barrack Street on the city side of Perth Water, is the Bell Tower, built in 1998, although the original concept plans were for a larger, more ambitious development. Here is found a small commercial development, including a bar, restaurant, and café development; however, it is bifurcated by a bus and car drop-off point, forcing pedestrians to contend with vehicles in a relatively confined space. Commercial ferry tours leave from the Barrack Street jetty, but with the exception of the South Perth ferry terminal, no other jetty structures offer amenities. In fact, Perth Water has very few jetties, and the majority of these are short, with no boating facilities. The permanent mooring of boats is not encouraged, although a number of short-stay and casual boat moorings can be found.

A shared cycle and footpath network traverses around Perth Water. At Point Fraser, tourists can rent bicycles and take short helicopter tours. The South Perth side provides tourists with considerably better access to the river and more destination and activity options. The public ferry terminates at Mends Street, which is an old-fashioned suburban main street with a variety of shops, cafés, and services in a leafy setting. Taking a walk along the river's edge east, a tourist has the opportunity to hire a catamaran, small yacht, or kayak, although no facilities are available for the tourist to shower or freshen up after a day on the river. Unfortunately, the walking paths lack interpretation of Aboriginal, natural, and historic sites, which could provide an interesting trail experience. The Perth Water walking and cycle paths could link with the extended foreshore trail network, but important parts of the path at East Perth are absent. This is a missed opportunity to add more than 20 kilometers to the foreshore trail and expand the benefits of tourism.

During the summer, occasional concerts take place, and on Australia Day each year, a spectacular fireworks display is held on the Swan River. Since 2006, Perth has hosted the Red Bull Air Race series, but this has recently been terminated. It is reported that 9,000 extra visitors came to Perth because of this air race, and it contributed more than A$14 million to the local economy.[2] All of these events were very popular with Western Australians, but residents who live close to the Swan River complained bitterly about the disruption and noise these events create.

In sum, very few events are held on or around Perth Water, and there is a dearth of variety. Tourists have limited opportunity to eat a meal or linger over a drink, and Perth Water has no galleries, few retail outlets, few boats of any size or description, and a lack of activities for visitors to interact with the Swan River. As noted in a personal communication by a senior local government executive who wishes to remain anonymous: "The Swan River would have to be regarded as one of the city of Perth's greatest assets, but it is also acknowledged that it is greatly underutilized. It seems clear that there is significant opportunity to enhance and promote the river for recreational and tourism use, but there appears to be no priority or vision by the state to increase the use of the river."

It is worth speculating on this statement. Who or what is thwarting better use of the Swan River for recreational and tourist use? Why is connectivity by, across, and through the river so limited? Who or what needs to change, and does Perth really desire something different?

ACCESS TO PERTH WATER

Today the banks of the river on the Perth side are almost exclusively hard, vertical walls, but this was not always the case. The city's river frontage is entirely human-made and extends 500 meters farther into the river than was the case when colonists first arrived in 1829. Then "the river margins were very shallow and edged with rush beds" (Seddon and Ravine 1986, 76). The first reclamation of the river was undertaken in 1903, when land about 250 meters wide and 3 kilometers long was reclaimed from the river with the construction of the extensive limestone river wall, culminating in what is now referred to as the Barrack Street jetties. The next large-scale reclamation occurred in the 1950s in preparation for the construction of the Narrows Bridge interchange, a span that joined Mount Eliza and the city to South Perth in 1959. As Figure 8.2 shows, those looking at the view from Mount Eliza to the city, Swan River, and Darling Scarp beyond must first look through a complex freeway interchange that in 2007 had a railway line built in the middle of it, thus reinforcing the pedestrian barrier and limiting connectivity between Mount Eliza, Kings Park, the CBD, and the river.

Seddon (1970) comments that the construction of the Narrows Bridge was the first, and decisive, act of obstruction of pedestrian access to the waterfront for the whole of its length, and this has been exacerbated by the construction of more roads and now a railway. A pedestrian wishing to access the river must dodge cars and traverse through a series of traffic lights. For those in a vehicle, the main road adjacent to the river has no viewing points where a car can pull off so that its occupants can admire the view. Also lacking is public transportation that will take a tourist along the river's edge on the city side.

Although in the 19th century, the river was an important transport network and ferries were the only way of linking some communities, the Swan River is now almost redundant as a transport conduit, notwithstanding the increasing traffic congestion. Figure 8.2 shows the Swan River on a spring afternoon; there is no sense of busyness, with no boats permanently moored or marinas with boats and people coming and going. Opportunities to travel on the Swan River are limited to one public passenger ferry service to and from South Perth, several private ferry services taking visitors up the river to the Swan Valley, and other, bigger ferries carrying patrons to Fremantle or Rottnest Island, 20 kilometers off the coast. The car is king, as shown in Figures 8.2 and 8.3. Attempts have been made to reintroduce an

Figure 8.2 *The City of Perth and the Swan River*

expanded passenger ferry service, but to date they have failed, primarily as a result of cost and low density of development in most riverside locations.

DULLSVILLE AND THE NANNY STATE

Kennewell and Shaw (2008) and others (CCI 2008; Demographia 2008; Edwards et al. 2007; Staley 2007; VCEC 2008) have shown that a city's profile, livability status, and reputation are important for attracting visitors and therefore for stimulating the local economy. Livability is highly valued for nurturing knowledge-intensive capitalism, competitive business investment, and the attraction and retention of a creative and skilled workforce(Florida 2005; Pinnegar et al. 2008). The more vibrant a city, the greater the likelihood of people staying longer and spending more while they are visiting. Research (Breen and Rigby 1996; Dodson and Kilian 1998; Sandercock and Dovey 2002; Stevens 2003) has shown that "waterfront redevelopment has gone hand-in-hand with a broader renaissance of inner cities … urban waterfronts have become key drawcards for foreign tourists, visitors from the suburbs and new upmarket residents" (Stevens 2003, 3). The Swan River, particularly Perth Water, therefore has an important role in contributing to the vibrancy and attractiveness of Perth.

A number of different schemes and plans have been submitted over the years to revitalize the Perth waterfront, from the Narrows Bridge to the Causeway, and to

Figure 8.3 *The Kwinana freeway and Perth to Mandurah Railway*

enhance access to the Swan River from the city. However, because of cost shifting state government priorities or conflicts between the state government (the owners of the land) and the city of Perth (the custodians), the status quo remains. Plans for private developments have been presented and some approved (including a riverside hotel and restaurant complex at Point Fraser) but not actually developed, because indecision has been too great regarding future foreshore developments, timelines, projected road changes, and public policy. Maginn identifies "a strong social and political conservatism that prevents urban planners, public and private alike, and developers from taking risks and pushing the boundaries of creating more dynamic, vibrant and off-the-wall inner-urban environments and thus experiences" (2010, 40). The local media and think tanks such as the Committee for Perth regularly decry the plethora of regulations that limit development and new ideas, describing policy- and decisionmakers as architects of a Western Australian "nanny state."

In 2007, FORM, a not-for-profit arts organization, conducted an online survey inviting the people of Perth to identify the challenges to the city's vibrancy and comment on its continuing ability to attract people. Fifteen key areas emerged from 2,681 responses, including the city's atmosphere or ambience, regulatory environment, leadership, lack of vision, "Dullsville" reputation, and need for more foreshore development, public facilities, and tourist attractions. The lack of world-class development on and connection with the Swan River was identified as a major issue. The feedback on regulation, mindset, and lack of vision was "remarkably consistent" (CCI 2008, 38). Respondents decried the numerous "nanny state" regulations and restrictions, the difficulty in getting development projects up and running, and the general sense of complacency that because Perth is located in a beautiful setting, it is therefore presumed "good enough" just as it is and hence attractive to visitors.

Submissions have once again been called for riverfront commercial development, and it would appear that both the state government and the city of Perth are in agreement with the general concepts presented, including increasing commercial, retail, and residential activity and facilitating more boating activity with protected bays, moorings, and jetty structures (Driscoll and Pearce 2008). An indigenous art and cultural center is a centerpiece of the most recent proposal, with the intention of providing a connection to indigenous peoples worldwide. The proposed center has the potential both to provide the opportunity for interpretation of the significance of the river system to Aboriginal people and to promote river and foreshore experiences to visitors and residents through Aboriginal cultural tours and art installations. The current Barnett government initially proclaimed that the redevelopment of the Perth foreshore would commence in 2011, but again priorities have shifted, and it is unclear if and when redevelopment will indeed go ahead.

The international planning consultant, urban strategist, and provocateur Charles Landry was invited to Perth in 2006 and 2007 to spark ideas and stimulate debate in response to a prolonged campaign by some Western Australians who claim that Perth is "Dullsville" (Kennewell and Shaw 2008; Staley 2007). Landry's first impression of Perth was "the plethora of signs prohibiting almost every conceivable street activity, ... mirroring a mind set of huge inertia, a culture of fear and risk aversion among Perth's bureaucratic leaders" (Laurie 2007, 15). Landry was reported to have been puzzled and frustrated that one of Australia's wealthiest cities "said no" with such frequency. Landry advocates "the need to embrace Perth's pre-urban indigenous heritage ... and to provide greater connections to Perth's hitherto jealously protected river foreshores" (Kennewell and Shaw 2008, 252). Similarly, 15 years after initially delivering a report on Perth's public spaces and public life, the Danish urban architect Jan Gehl was invited to return to Perth in 2009 to assess whether Perth had indeed "improved" as a place for people. In the 2009 report, he is again critical that "the fabulous setting of Perth is under-utilised. The proximity to the Swan River can hardly be perceived when in the city centre. The park landscaping along the foreshore, as well as the river itself, offer few activities and access to the water's edge is poor ... Kings Park is, despite the proximity, more or less inaccessible by foot from the city centre due to the barrier effect of the freeway" (Gehl Architects 2009, 14). Gehl recommended in 1994 and again in 2009 that the river foreshore be better connected to the city areas, enhancing street vistas to the Swan River, and that a plethora of low- to medium-rise mixed retail, commercial, cultural, and residential buildings be constructed, thus encouraging a sense of busyness, variety, and surprise by using the city's greatest natural asset, the Swan River, more creatively.

Favorable international business conditions for much of the last decade have prompted substantial investment in Western Australia, but despite resource extraction being a key driver of the Western Australian economy, economic activity is largely urban-focused and urban-directed. Pinnegar et al. (2008, 19) note that the "higher order functions are city-focused" and the knowledge-intensive services and key growth job sectors require city locations. The nature of the investment demands creative, entrepreneurial, skilled workers who operate in a highly mobile global market and are used to highly competitive, vibrant cities. The Chamber of Commerce and Industry Western Australia describes the situation: "The world is becoming a place where skilful workers are determining their employment conditions, high amongst which is location. Low unemployment rates … ageing workforces and increasing specialisation are making some types of employees price makers rather than price takers" (CCI 2008, 12). It is imperative that Perth differentiate itself, building on its considerable attributes—including its subtropical Mediterranean climate, proximity to Asia, strong economy, and unique natural resources—to enhance its international status as a highly livable and vibrant city, not only to attract and maintain a local workforce, but also to draw tourists visiting for varying periods of time.

SWAN RIVER GOVERNANCE

Although the Swan River is a beautiful natural asset, it is in environmental trouble and has been for some time. Until a little more than a decade ago, commercial fishermen operated in the river, but because of pollutants, fish stocks are depleted and the fishermen have now gone. In 2009, dolphins were reported to have died inexplicably in the river, and after examination, it appears that the cause of death was long-term exposure to contaminants and chemical buildup. In 2004, the premier pronounced the Swan River to be Western Australia's first official heritage icon and committed an additional A$15 million over the next four years. This was to be spent on restoring river foreshores, limiting nutrient flow, and establishing a new Swan River Park and Swan River Act. These acts gave greater powers to the Swan River Trust, the agency charged with the protection of the river (Constitutional Centre of Western Australia 2008). Five years later, telltale signs showed that the waterway was continuing to deteriorate and only limited new development had occurred on the river, despite numerous task forces, submissions, master plans, and peer reviews (Driscoll and Pearce 2008). People claim that the river has been "locked up."

We conducted interviews with a variety of stakeholders who have a role in the governance of the Swan River, undertake business on or around the river, or promote tourism in the CBD. The overwhelming sentiments were that the Swan River was an undervalued natural asset, particularly for tourism purposes. Interviewees expressed frustration that despite people's and organizations' intentions, little has changed, even in spite of a parade of international thinkers, urban strategists, and local commentators urging government, the community, and business to "do something." All agreed that the Swan River requires significant funds for maintenance of current infrastructure and rehabilitation of the riparian environment, and that current budgets do not come close to rectifying the damage.

Moreover, confusion is widespread about which agency or organization is responsible for the provision of funds, liability, and ratification of particular decisions. "This, against an historic backdrop where different agencies, government departments and local governments have had varying levels of responsibility for the maintenance of different sections of the foreshore, has resulted in "non-action" and even court action between levels of government to determine liability" (City of South Perth 2008, 3).

Local government authorities and regional councils recognize the value of tourism to their constituents and businesses, but because of budgetary constraints, their efforts seem doomed to be piecemeal projects that are inevitably uncoordinated and create service gaps. From a business and investor perspective, the numerous town-planning schemes and multitude of local bylaws vested in so many local government authorities act as a barrier to development and business innovation. Further, as Stilwell and Troy (2000) note, decisions by local government authorities regarding development applications often are not compatible with the broader strategic plans of the state.

The role of the Swan River Trust, a state government agency responsible to the minister for the environment, appears to be misunderstood by a variety of agencies and organizations at every level of the community. A general misconception exists that the trust has operational responsibilities, when in fact all it does on a very small budget is provide advice regarding facilities, environmental health management, protection, rehabilitation, and town-planning issues. Funds are allocated by the state through the Swan River Trust to local government and other jurisdictions with operational responsibility, and each year for the last three, the amount has decreased. In the 2009–2010 fiscal year, the budget to be shared among the 21 local government authorities with operational responsibilities for maintenance and improvement works was A$600,000, down from A$1 million the year before. This is in spite of the trust having identified in 2008 priority works along the Swan and Canning Rivers totaling A$82 million from a total works schedule of approximately A$225 million (City of South Perth 2009). It is not surprising that those organizations with operational responsibility are frustrated. The local government sector, in particular, is concerned that the expenditure required to reach the full potential of the Swan River is well beyond its capabilities, and that it is constantly the victim of cost shifting. Furthermore, many of the 21 local government authorities with frontage on the Swan and Canning Rivers are small and lack capacity to perform major capital works of any description, let alone large-scale infrastructure works that have a wider regional benefit.

Private enterprises regularly complain that it is hard to do business on the river. The plethora of agency approvals required and the time taken heighten investment risk and undermine profitability. A development on the Swan River may require approval from some 10 government agencies, depending on the size and nature of the development. Developers complain that they cannot apply for all approvals simultaneously; approval from one agency is a prerequisite for approval from another. Bureaucrats also become frustrated. One inter-viewee stated that "doing anything on the river as a bureaucrat is a potentially career-making and career-breaking exercise, depending upon your level of patience." The Western Australian Chamber of Commerce and Industry has regularly complained that poor coordination among the different levels of government inhibits business investment and a multiplicity of regulatory regimes snuffs out creative initiatives and entrepreneurialism.

Any development on, in, or beside the river also requires clearance in accordance with Section 18 of the Aboriginal Heritage Act 1972. The traditional owners of a particular place must be properly consulted about a proposed development and have input to the minister for indigenous affairs as to whether consent should be given to use the land (or water) as proposed, based on Aboriginal heritage values. Seeking input from Aboriginal stakeholders requires time and cultural understanding, and the process is often made even more difficult by sometimes fraught Nyoongar politics. In addition, the Environmental Protection Act 1986 takes precedence, although not all development proposals are required to have Environmental Protection Authority referral.

The upshot is that confused hierarchical approval and referral arrangements cause confusion and duplication, often leading to litigious conflict, frustration, and inevitable delays in decisionmaking.

On several occasions, interviewees noted that although it was easy to blame bureaucrats and government more generally for the lack of development and access to the river, another problem was NIMBYism, which had been allowed to proliferate in Perth and had effectively stultified innovation and development. Staley (2007) is critical of the heavy-handed approach to regulations and the ease by which applications can be rejected based on a small number of complaints or a vociferous minority. da Silva (2008), on the other hand, observes that the Western Australian public expects too much of its public bodies and regulators "when the crowd goes mad." He argues that "the behavior of crowds is not always rational" and cannot and should not be regulated by a governance model.

Sweeping reforms to the development approvals process have been recommended. This would see planning approvals largely removed from the local council's jurisdiction and decided by a panel including two experts, two local councilors, and a government-appointed chairperson (DPI 2009). This has been warmly welcomed by the development and building industries, but the local government sector has vigorously attacked the principles of such reforms as an intrusion into local rights and responsibilities, underscoring the tensions identified by Stilwell and Troy (2000) and others (Devine-Wright 2009; Malpezzi 2001) between the state and the community. Malpezzi's work (2001) shows that the more stringent the regulatory environment, the higher the housing prices and the slower the growth. It is not surprising, then, that local government is accused of preserving community interests and the status quo; Perth property prices have become some of the most expensive in the country (ABS 2008; HIFG 2009). NIMBYism is a pervasive barrier to "doing something" (Devine-Wright 2009), particularly in and around Perth Water, where property prices are at a premium.

HARLEM RIVER PARK, NEW YORK

NIMBYism and complacency are not unique to Perth, of course. The revitalization of New York City's Harlem River waterfront came after many years of neglect and inertia. The Harlem River divides the island of Manhattan from the Bronx. The city's waterfronts were taken for granted, and large expanses of waterfront languished as abandoned industrial wasteland in the face of a general unwillingness to challenge traditional practices (Wagner 1980). This was changed when a visionary mayor, Michael Bloomberg, together with a number of community, corporate, and public organizations, began to drive innovative revitalization programs throughout the city, with the PlaNYC, a sustainability program for the city's future, being released on Earth Day 2007 (Platt 2009).

The revitalization of Harlem River Park was part of New York City's "gradual transformation of its long-degraded waterfronts for new uses and users" (Platt 2009, 48). By the 1970s and 1980s, this postindustrial city no longer had any use for the unsightly structures that had been built in a bygone era to facilitate river transport for the industrial districts on the Harlem River. Nowadays, design responses for reuse of the waterways in line with a changing cityscape more oriented to visual amenity and tourism have been adopted (Savitch 2010). Until recently, the standard waterfront treatment has been "a paved esplanade with an ornamental steel railing adjacent to a vertical masonry or steel sheetwall" (HRPTF 2006, 1), but contemporary attempts at waterfront redevelopment in Manhattan have assumed a more nuanced approach. Changes began in 1992 under Mayor David Dinkins, with a plan envisaging a 21st-century waterfront that facilitated recreation, tourism, and economic development for the Harlem River Park area (NYC DCP 1992). In 2002, Mayor Michael Bloomberg put forth a new revitalization program that envisioned the following scenario on the Manhattan waterfront (City of New York 2002):

- parks and open spaces with a lively mix of activities within easy reach of communities throughout the city;
- people swimming, fishing, and boating in clean waters;
- restored, well-cared-for natural habitats;
- maritime and other industries, though reduced in size from their heyday, thriving in locations with adequate infrastructure support;
- ferries crisscrossing the city's harbor and rivers and interconnected systems of bikeways and pedestrian pathways, all helping lessen traffic congestion and air pollution;
- panoramic water views of great beauty; and
- new housing and jobs for people of diverse incomes.

Consolidating the principles established earlier, the Manhattan Waterfront Greenway Master Plan, approved in 2004, outlined an extended vision that included a role for tourism and a "continuous shared use pedestrian and bicycle path around Manhattan to enhance recreational opportunities for all New Yorkers and provide a green attraction for those from outside of the City" (NYC DPR 2004, 1).

The Harlem River Park Task Force, the overarching body responsible for Harlem River Park development, acknowledged that waterfront redevelopment responses in Manhattan represented satisfactory initial solutions, while also recognizing that the public now demanded more: "The demand for public access [to waterfront areas] includes ferries for transportation, tourism and recreation; emergency evacuation by boat; launches for kayaks, rowboats and canoes; and research or educational access to the intertidal zone where crabs, shrimp, shellfish and other aquatic life can be observed on the edges of the NY/NJ [New York/New Jersey] Harbor estuary" (HRPTF 2006, 1).

The design process for Harlem River Park centered on a philosophy of engagement with the full range of stakeholders, especially community and professional groups. The task force conducted meetings with Manhattan community boards and their parks and cultural committees "before convening a successful community design charrette to gather community input." The charrette incorporated all stakeholders, including a New York City Parks and Recreation Department waterfront specialist, landscape architects and planners, a marine engineer, a marine biologist, environmental artists, community board members, representatives of elected officials, community organizations, tenant groups, and residents of Harlem, all of whom participated in the final design process (HRPTF 2006, 1).

The design phase teased out critical accessibility issues, including one major impediment to access: Harlem River Drive, an expansive, six-lane north-south highway. Pedestrian access to the waterfront is basic and purely functional, achieved via a walkway, a pedestrian bridge at 142nd street, or the Madison Avenue Bridge (HRPTF 2006). The site is linked with the wider Manhattan region by a continuous pedestrian and bicycle path, slowly being constructed around the periphery of Manhattan (see NYC DPR 2004). Accessibility is further enhanced by public transport, with high-frequency subway services on two subway lines and regular east-west bus services (HRPTF 2006).

The site has been developed in three phases by the New York City Department of Parks and Recreation. Phase I opened in November 2002, and the remaining two phases were opened to the public in the summer of 2009 (HRPTF 2006; HCDC 2009; NYC DPR 2009). The second and third phases of redevelopment wholeheartedly embrace the waterfront concept, providing a diverse range of activities and amenities at waterfront sites. The redeveloped river edge at Harlem River Park sought to include a diversity of experience and amenities. The Harlem River Park Task Force (2006, 1) outlined the following design features:

- surfaces that support estuarine life, which encourage algae growth and provide habitats for different types of aquatic life;
- seawalls with irregular shapes to enhance interest and reduce erosion from fast-flowing water;
- greenery and biomediation earthen banks to filter rainwater runoff from polluted areas;
- hand-powered boats such as kayaks and rowboats to encourage diversity and limit shore damage;

- the capacity for visiting and larger boats and emergency vessels; and
- safe access incorporating design solutions that allow people to safely interact with the water.

With such a large extent of river waterfront throughout the five boroughs of New York in need of redevelopment, the city's experience has shown that early solutions, which generally conformed to a standard template providing a paved area and security railing, were not the variety of development desired by the residential and business community. Existing projects, most notably Harlem River Park, have engaged the community in the design process and consequently provided an amenity that is both attractive and engaging. This type of process has continued throughout the city, with similar community consultation and design solutions (NYC DPR 2009; Platt 2009).

The New York City waterways revitalization experience provides some useful lessons for decisionmakers and residents of Perth. Just as is the case in Perth Water now, decades of wrangling by New York City bureaucrats, railroad companies, neighborhood residents, and commuters over the waterfront meant that the contested real estate languished, marine life suffered, and mediocre planning responses continued, frustrating everyone involved. Community pressure in the 1980s galvanized strong civic leadership to take action. A comprehensive design process provided the overarching guidelines necessary to oversee construction of the diverse components. The key features emphasized the importance of accessibility, recreation, diversity of use, and ecological improvement, all of which have also been identified as important in the future of the Swan River. The inclusion of a wide range of participants in the decisionmaking and planning of the revitalization programs was important for buy-in of the redevelopments and long-term investment in public spaces and assets. Rather than suspicion between the private and public sectors, as is the case in Perth, those involved in the Harlem River Park revitalization recognized that the interests of all sectors had to be balanced, and that without wealth creation, the long-term future success of the area was not possible.

CONCLUSIONS

The Swan River at Perth Water is a stunning stretch of water that is important to Western Australians and claimed to be a significant tourism attraction. This chapter has argued, however, that although the Swan River is an asset that many tourists admire, few actually experience or interact with it, and signs indicate that all is not well with this body of water. Examination of the tourism options on Perth Water indicate that a visitor's experience of the Swan River is constrained by poor accessibility and connectivity to other key sites along and beside the river. At the same time, residents of Perth despair of their city's "nanny state" and "Dullsville" reputations and implore that somebody do something about the situation. The ecosystems are struggling, suggesting that embedded environmental problems in the waterway have not been adequately addressed. The lack of development on the river is indicative of both complexity and complacency.

Because the Western Australian economy is experiencing phenomenal growth, it is tempting to dismiss the importance of tourism and its contribution to the broader economy. However, global competition for skilled labor has shown that Western Australian industry is vulnerable, and unless Perth can demonstrate vibrancy, livability, and a high ranking among other international cities, many mobile, discerning workers will choose to live and work in other cities. Refining the tourism experience is an important indicator of broader livability.

Government and good governance have vital roles. Unambiguous governance structures that facilitate timely and informed decisionmaking, enabling government with a clear, overarching vision and leadership, are of critical importance. Pinnegar et al. (2008) show that cities with effective governance structures are using their strengths to shape and drive expectations to ensure competitive advantage.

After many false starts, the Perth waterfront must be revitalized in the near future. It is critical that the river's cultural heritage, beauty, and ecology are celebrated by both visitors and residents so that, just as has been the case in New York, the people of Perth value all that it has to offer and willingly invest in its rehabilitation and preservation. There is no room for complacency.

NOTES

1. NIMBY is an acronym for "not in my backyard," referring to people who oppose projects for development because of the perceived effects on their own quality of life, property values, or both.

2. A$1 = US$0.9978 as of January 2011.

REFERENCES

ABS (Australian Bureau of Statistics). 2008. *House Price Indexes: Eight Capital Cities.* cat. 6416.0. Canberra: Australian Government Publishing Service.

Breen, A., and D. Rigby. 1996. *The New Waterfront: A Worldwide Urban Success Story.* London: Thames and Hudson.

CCI (Chamber of Commerce and Industry Western Australia). 2008. *Perth Vibrancy and Regional Liveability: A Discussion Paper.* Perth: Chamber of Commerce and Industry Western Australia.

City of New York. 2002. *The New Waterfront Revitalization Program.* DCP #02-14 (September), accessed January 23, 2011, from www.nyc.gov/html/dcp/pdf/wrp/wrp_full.pdf.

City of Perth Council. 2009. *Point Fraser,* accessed January 22, 2011, from www.perth.wa.gov.au/web/Council/Environment/Point-Fraser/.

City of South Perth. 2008. *Infrastructure Australia Business Case: Impacts of Climate Change on Swan and Canning River Foreshores.* South Perth: City of South Perth.

City of South Perth. 2009. CEO, City of South Perth, personal communication with the author.

Constitutional Centre of Western Australia. 2008. *Heritage Icons: The Swan River,* accessed October 22, 2009, from www.ccentre.wa.gov.au.

da Silva, R. 2008. Corporate governance delusions and the madness of crowds. In *Corporate Governance (Breakfast & Seminar).* Perth: Institute of Public Administration Australia (IPAA) WA Division.

Demographia. 2008. *4th Annual Demographia International Housing Affordability Survey*, accessed March 25, 2008, from www.demographia.com/dhi.pdf.

Devine-Wright, P. 2009. Rethinking NIMBYism: The role of place attachment and place identity in explaining place protective action. *Journal of Applied Social Psychology* 19: 426–441.

Dodson, B., and D. Kilian. 1998. From port to playground: The redevelopment of the Victoria and Albert Waterfront Cape Town, in *Managing Tourism in Cities*, edited by D. Tyler, Y. Guerrier, and M. Robertson. Chichester: John Wiley, 139–162.

DPI (Department of Planning and Infrastructure). 2009. *Building a Better Planning System: Consultation Paper*. Perth: Department of Planning and Infrastructure.

Driscoll, P., and D. Pearce, (Eds.). 2008. *The Shaping of the Perth Waterfront Masterplan*. Perth: Western Planner.

Edwards, D., T. Griffin, and B. Hayllar. 2007. *Development of an Australian Urban Tourism Research Agenda*. Brisbane: Sustainable Tourism Co-operative Research Centre.

Florida, R. 2005. *Cities and the Creative Class*. New York: Routledge.

Gehl Architects. 2009. *Perth 2009: Public Spaces and Public Life*. Perth: City of Perth and the Department of Planning and Infrastructure.

HCDC (Harlem Community Development Corporation). 2009. *Harlem River Park Task Force*, accessed November 18, 2009, from www.harlemcdc.org/planning/planning_hr_park.htm.

HIFG (Housing Industry Forecasting Group). 2009. *Dwelling Commencement in Western Australia*. Perth: Housing Industry Forecasting Group.

HRPTF (Harlem River Park Task Force). 2006. Designing the edge: Where land and water meet. *The Edge News* accessed November 18, 2009, from www.harlemriverpark.com/edge_news.pdf.

Kennewell, C., and B. Shaw. 2008. City profile: Perth. *Cities* 25 (4): 243–255.

Laurie, V. 2007. Shaking Up Dullsville. *The Australian* March 15.

Maginn, P. 2010. Conservatism stifles waterfront evolution. *WA Business News* :40 February 18.

Malpezzi, S. 2001. *NIMBYs and Knowledge: Urban Regulation and the "New Economy"*. Berkeley, CA: Housing and Urban Policy, Institute of Business and Economic Research.

Marzano, G., E. Laws, and N. Scott. 2009. 'The River City'? Conflicts in the development of a tourism destination brand for Brisbane, in *Water Tourism*, edited by B. Prideaux and M. Cooper. Wallingford, UK: CAB International, 239–256.

NYC DCP (New York City Department of City Planning). 1992. *New York City Comprehensive Waterfront Plan: Reclaiming the City's Edge*. New York: Office of Mayor Dinkins and New York City Department of City Planning.

NYC DPR (New York City Department of Parks and Recreation). 2009. *Greenpoint-Williamsburg Waterfront*, accessed November 18, 2009, from www.nycgovparks.org.

———. 2004. *Greenpoint-Williamsburg Waterfront*, accessed November 18, 2009, from www.nycgovparks.org.

Pinnegar, S., J. Marceau, and B. Randolph. 2008. *Innovation and the City: Challenges for the Built Environment Industry*. Sydney: City Futures Research Centre, University of New South Wales.

Platt, R. 2009. The humane megacity: Transforming New York's waterfront. *Environment* 51:46–59.

Sandercock, L., and K. Dovey. 2002. Pleasure, politics and the public interest: Melbourne's waterfront revitalisation. *Journal of the American Planning Association* 68 (2): 151–164.

Savitch, H. 2010. What makes a great city great? An American perspective. *Cities* 27: 42–49.

Seddon, G. 1970. *Swan River Landscapes*. Perth: University of Western Australia Press.

Seddon, G., and D. Ravine. 1986. *A City and Its Setting: Images of Perth, Western Australia*. Perth: Fremantle Arts Centre Press.

Staley, L. 2007. *Creating a Liveable City*. Melbourne: Institute of Public Affairs and Mannkal Economic Education Foundation.

Stevens, Q. 2003. *Australian Waterfronts: Improving Our Edge*. State of Australian Cities National Conference Proceedings, Brisbane.

Stilwell, F., and P. Troy. 2000. Multilevel governance and urban development in Australia. *Urban Studies* 37 (5-6): 909–930.

Tourism Queensland. 2009. *Queensland Data Sheet*. Brisbane: Tourism Queensland.

Tourism Western Australia. 2009. *Overnight Visitor Fact Sheet 2007/2008/2009*. Perth: Tourism Western Australia.

Urbis. 2008. *Heirisson Island Sculpture Park Masterplan*. Perth: Urbis.

VCEC (Victorian Competition and Efficiency Commission). 2008. *A State of Liveability: An Enquiry into Enhancing Victoria's Liveability*. Melbourne: Victorian Competition and Efficiency Commission.

Wagner, R. Jr. 1980. *New York City Waterfront: Changing Land Use and Prospects for Redevelopment*. Washington, DC: National Academy of Sciences.

CHAPTER 9

Recreational Access to Urban Water Supplies

Michael Hughes and Colin Ingram

C oncerns over the global capacity to support a growing world population and cope with a changing climate has resulted in a change of focus regarding the management of the world's water resources over the past few decades (Pahl-Wostl et al. 2007; Pigram 2006). For example, the management of water in Europe has shifted from a primary focus on public health and standards for drinking water in the 1970s and 1980s, to water pollution control and environmental management in the late 1980s to mid-1990s, and then to strengthening legislation to protect water resources through integration to meet new scientific knowledge and associated legal obligations by the turn of the 21st century (Page and Kaika 2003). For example, recreational use of drinking-water dams and catchments is actively promoted in the United Kingdom. Consequently, on an international level, provision for tourism and recreation opportunities is commonly incorporated into water storages for irrigation and drinking-water supply. It is important that the legislative, policy, and planning frameworks are integrated and capable of catering to these multiple uses (Ingram 2009).

As discussed in detail in Chapter 6, many perspectives exist on what constitutes good governance (see also UN 2009). Chapter 6 draws on evidence of a global shift in environmental governance toward more collaborative approaches, and Chapter 11 takes up and further develops this theme. Research on the trends in protected-area governance undertaken for the 2003 World Parks Congress reveals a major movement toward an inclusive and participatory approach to protected-area planning and management (Dearden et al. 2005). This is evident in the United Kingdom, United States, and across much of Western Europe, where governments are attempting to shift the focus toward various forms of comanagement with other agencies and with the community through partnerships and public involvement (Newman et al. 2004).

In contrast, in Western Australia (WA), the predominant response to complex issues of water catchment management has been a regime of exclusion. This raises serious concerns, given the overwhelming evidence that recreation in water catchments and dams holds social value through improved public health, quality of life, and stronger community networks (Martinick & Associates 1991). Research at Deakin University indicates that people engaged in nature recreation develop or reinforce positive social networks and improve personal well-being, including physical and mental health (HCN/RMNO 2004; Maller et al. 2008; Natural England Board 2007; Sharp 2005). The loss of recreation and tourism opportunities extends farther, having direct and indirect economic impacts.

This chapter examines the complex issues associated with urban water catchment management incorporating recreation and tourism, using WA as a case reference. The current pressure on WA's water resources is severe and likely to increase. Because of WA's generally dry climate, water bodies that afford adequate volumes of water for drinking, irrigation, and recreational use are particularly valued in the southwest of WA (Muench 2001). Thus, the management of water for water supply alone is problematic. The associated governance of water is even more complex, involving regulation and policy considerations with social, environmental, and economic dimensions at the local, state, national, and in some cases international levels. This is further complicated by water policy considerations for other uses, such as industry and agriculture, which have entrenched property rights and secure access through volumetric purchases. Access to water for nonconsumptive use is extremely difficult to secure within a volumetric, market-based water resource system (Ingram 2009). Recognition of these multiple uses and values is essential for sustainable management of water catchments and the resulting benefits to all users.

In this chapter, we address past and current water management problems for WA and propose a model for more effective governance. We first examine the Western Australian context, considering the state's history, types of recreational activity, and future challenges. Following this, we turn our attention to the problems of management, policy, and legislation. We then draw on the Victorian experience to provide some lessons for WA.

THE WESTERN AUSTRALIAN CONTEXT

The majority of the current WA population of 2.5 million lives in the temperate southwest region of the state. Water catchments have historically been an important part of the recreation experience in this region (Martinick & Associates 1991). The southwest water catchments are located within state forest and timber reserves located along the Darling Range (Figure 9.1). The Darling Range area includes about 3 million hectares of state forest (timber reserve) and a further 1.1 million hectares in existing and proposed national parks and other conservation reserves (CCWA 2003; DEC 2004). The Darling Range lies adjacent to the state capital and major urban corridor of Perth and other urban centers. The metropolitan area of Perth extends along a coastal plain, situated between the Indian Ocean and west of

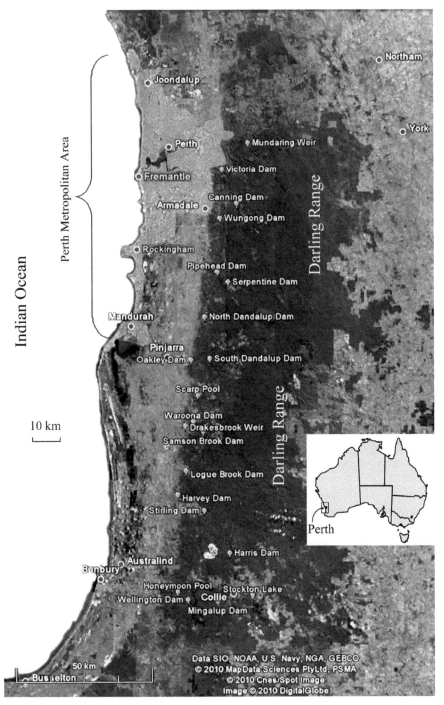

Figure 9.1 *Map of Darling Range and location of urban areas and dams*

the Darling Range, for 120 kilometers north to south and about 50 kilometers east to west. Of the state's total population, 1.5 million currently live within the greater metropolitan area of Perth. In addition to drinking-water catchments, the Darling Range functions as a major recreation area and provides resources and space for various agricultural and industrial activities, such as cropping, grazing, mining, and timber, as well as towns and residential areas.

History

The Nyoongar Aboriginal people occupied and sustainably managed the southwest region of WA for at least 45,000 years prior to European colonization. Water and water catchments are important physical, spiritual, and cultural elements in the lives of Aboriginal people. Fresh water is relatively scarce in Australia, and understanding of water catchments and water cycles was essential for survival of Aboriginal family groups prior to the establishment, expansion, and urbanization of European colonies. Consequently, knowledge and use of water catchments were and are central to the heritage and culture of the Nyoongar occupants of the WA southwest region. The British colonized the southwest of Western Australia with a military outpost established on the south coast at Albany in 1826 and a subsequent colony at the Swan River in 1829, which later became the city of Perth. Although the Nyoongar lived in relative but uneasy peace with the early settlers, the ongoing influx of colonists, expansion of occupied land, reduction in access to food sources, and resultant displacement of Aboriginal families soon led to conflict. Colonists responded to Nyoongar raids on farms to obtain food and supplies with violent retribution and officially condoned massacres of Aboriginal family groups. The British colonization of the southwest WA region severely disrupted the Nyoongar way of life, resulting in a fragmentation of Aboriginal cultural structures and practices (SWALSC n.d.). Despite a history of being oppressed and marginalized, the Nyoongar have retained much of their cultural practice and knowledge. Today, contemporary programs are being developed and run by the Nyoongar to maintain their strong connections with water catchments and to foster cultural practices and a sense of identity.

In terms of the postcolonial period, recreation-based tourism has been occurring in water catchments in the southwest of WA for more than 150 years. The early colonial history of WA saw forestry reserves and national parks declared in the Darling Ranges during the early 20th century, primarily as a means of preserving timber for future forestry enterprise, as well as for recreation (Herath 2002). As the timber industry grew, the forested areas became more accessible, and recreation in forests became more dispersed and popular. Increased accessibility for a wider segment of the population resulted in a broader range of recreational pursuits and competitive sports (DCLM 1992; WRC 2003).

In 1955, an urban plan was created for Perth (Stephenson and Hepburn 1955). The plan formally addressed establishing recreation and conservation reserves for the public good, proposing that large areas of open space be set aside for these purposes, including water catchment areas on the Darling Escarpment. The intention was to preserve areas of natural bush land as well as provide spaces for people to escape the growing urban sprawl of Perth for recreation and rejuvenation (Weller 2009). The dams and associated catchments remained a significant focus for recreation, essentially endorsed by the 1955 plan. Recently, however, some WA state government agencies responsible for water management have begun to limit recreational access to catchments. This is seen as a means of preserving future drinking-water supplies for the Perth metropolitan area and represents a move away from the use of forested catchments as recreation resources.

Types of Recreation in Western Australia Water Catchments

A shift in recreation has occurred from "passive pursuits" in the early 1970s to a more diverse range of activities, including adventure-based pursuits, by the 1980s. Muench's 2001 study conducted in the Southern Darling Range provided a list of 13 popular types of recreational activities in southwest catchments: picnicking, designated camping, wild camping, "bushwalking" (hiking), trout fishing, "marroning" (catching large freshwater crayfish known as marrons), flat-water canoeing, whitewater canoeing, four-wheeling, swimming, water-skiing, sightseeing, and rock climbing.

Ingram and Hughes (2009) later documented 13 different physically active recreational activities (with some activities including several variations) undertaken by formal recreation clubs and associations within the Southern Darling Range of WA: "bushwalking," fishing, endurance horse racing, trail biking, mountain biking, four-wheeling, swimming, canoeing and kayaking, water-skiing, climbing, shooting, rogaining (long-distance cross-country navigation), and orienteering (navigating from point to point with the aid of map and compass). This list indicates diversity not just in terms of the types of recreation, but also within each recreation type in terms of people and their requirements.

These studies demonstrate that the WA water catchments are used for a wide variety of recreation and tourism-related purposes, and the water catchments in other Australian states as well as internationally experience a similar range of uses (Hughes et al. 2008). Despite a consistent growth in diversity and demand for recreational pursuits, areas for water-based recreation in WA are diminishing as a result of restrictions on recreation in water catchment areas, inevitably leading to capacity and access management issues elsewhere.

The Future for Western Australia

The climate of southwest WA is undergoing significant warming and drying, with reduced winter rainfall and increased temperatures associated with anthropogenic climate change. The increase in mean temperatures for the region has led to increased evaporation rates, and this along with reduced winter rainfall has resulted in a significant reduction in streamflow in the southwest of WA over the past two to three decades that is projected to continue into the future (Yates et al. 2010). These conditions will function to reduce the availability of surface-water bodies, both for urban supply and tourism and recreational uses. As freshwater availability declines, an increasing population means that demand for consumptive and nonconsumptive uses is likely to rise in the southwest region.

In 1984, the Tourism and Recreation Committee Report to the System 6 Study on the Darling Range region of WA estimated that demand for outdoor recreation would at least triple by the year 2000, based on a population increase that was estimated to double by 2021 due to accelerated urban growth in adjacent eastern and southeastern corridors (cited in Ingram and Hughes 2009). These predictions have proved to be reasonably accurate to date. A report by Feilman Planning Consultants (1987) attributed the growth in recreational demand during the 1980s to population growth, increased discretionary time and money, developments in transportation increasing mobility, discerning and educated individuals seeking different experiences, and the loss of opportunity for recreation in nonforested areas. The Feilman report predicted forest visitation in excess of 3 million visits per annum by 2010 and estimated that demand for outdoor recreation was increasing at a rate greater than the rate of population growth of Perth. The 2002 *Draft Forest Management Plan* (DCLM 2002) recorded 4.6 million visits to Department of Environment and Conservation managed southwest forest regions in 2000–2001. Present demand for recreation in southwest forests reached 2.4 million visits in 2008–2009 (DEC 2009). Recreational use of southwest forests (excluding coastal areas) has grown consistently at an average of just under 2% per annum since 1995–1996.

The population of the Perth metropolitan area is projected to reach 2.2 million by 2020 and more than double by 2050 to approximately 4.2 million people (Weller 2009). This is likely to result in ever-increasing urban sprawl along the coastal plain, with population centers, such as Rockingham, Mandurah, Bunbury, Busselton, and Dunsborough, hosting growing populations. People living in these urbanized environments use the Darling Range and its urban water catchments as recreational resources and for temporary escape from city life (Ingram and Hughes 2009). The continuing growth of the Perth metropolitan area will be accompanied by increasing demand for nature-based recreational and tourism experiences in water catchments, along with a growing need for consumptive uses, such as drinking water and agriculture). This has placed pressure on water catchment managers to recognize and make allowance for the multiple use requirements associated with WA water catchments. However, the existing legislative, policy, and management environment is resistant to changing the current "single-use" approach.

MANAGEMENT, POLICY, AND LEGISLATION IN WESTERN AUSTRALIAN CATCHMENTS

Since 1903, WA has experienced a steady decline in the number of water catchments accessible for recreation. For example, 7 of the 10 significant rivers in the upper southwest have been dammed over this time, with a consequential loss of recreational opportunities in many areas, including the Helena, Canning, Wungong, Serpentine, South and North Dandalup, Harvey, and Collie Rivers

(Thorpe 2006). This loss is a result of a management policy of exclusion of public recreational and tourism access based on water contamination concerns (Hughes et al. 2008).

Four main players govern the management of land and water catchment access in the southwest of WA. These include three state government agencies—the Department of Environment and Conservation, Department of Water, and Water Corporation—with the fourth player being the recreational users of the land. This fourth group includes many independent users, as well as various clubs and associations that have the capacity to lobby government and attempt to influence policy but also encourage responsible behavior among their members. Each of the four players has adopted a specific stance with regard to recreational access to drinking-water supply catchments.

Several acts mandate the various government agencies, bylaws, and regulations related to the management of drinking-water supply catchments. However, these often have contradictory requirements in terms of land management and public access. This is complicated by agencies having overlapping land jurisdictions and management responsibilities. It is often unclear which laws have primacy in these instances and therefore which agency takes priority in terms of management.

Water Catchment Management

To understand why recreational access to water catchments has been restricted in WA, it is necessary to understand the roles of the various organizations responsible for managing these areas. The Department of Environment and Conservation (DEC) has a statutory responsibility for managing large areas of the state, including water catchments, for a range of values. These include nature conservation, recreation, water catchment protection, and in the case of state forest, timber production. DEC is also mandated to facilitate public access to natural areas, including water catchments, for recreation and tourism. Consequently, DEC balances the public access requirements with water catchment protection based on a risk management approach. That is, DEC takes the position that risks to water quality arising from recreational activities can be adequately managed. This is based on the notion that allowing low-risk activities in sensitive areas combined with adequate facilities and an appropriate management presence can mitigate risks to water quality in catchments and the urban water supply (Hughes et al. 2008).

Risk management can allow certain types of low-risk recreation combined with protective measures based on a series of barriers. If one protective barrier fails, other barriers should be sufficient to compensate (DEC 2004). Multiple barriers in drinking-water catchments may consist of a combination of regulation and surveillance, physical barriers, appropriate form and location of recreation facilities and activities, water filtration, and treatment processes (where necessary), all of which together enhance the security of water quality. The multiple-barrier approach to catchment management is advocated for water supply areas where recreational access is allowed (Patterson 1977).

The Department of Water (DoW) has the function of safeguarding water quality and implementing control measures under powers granted to it by the water resource legislation. DoW approaches water catchment management from a "risk avoidance" stance, with any level and type of recreation considered a risk to water quality and therefore excluded from large buffer zones established around drinking-water dams. However, in some cases, historical activities or facilities are permitted, causing confusion and contradicting the intended policy outcome.

DoW's Policy 13 specifies classification levels, definitions, and a broad framework for public drinking-water supply protection and management. It is through this policy that DoW outlines its primary objective of ensuring quality drinking water and protection of drinking-water dams and catchments. This policy defines a three-tiered classification system for protection of water bodies and catchments (Hughes et al. 2008):

1. *Priority 1 (P1) Areas* generally occur on Crown land, protected areas, and state forest managed by DEC. They are demarcated to prevent degradation of the water source and are declared over land that DoW considers prime for providing the highest-quality public drinking water. DoW uses a "risk avoidance" approach for P1 areas, meaning that recreation is generally excluded.
2. *Priority 2 (P2) Areas* are designated over land where low-intensity development is taking place if DoW considers it necessary to ensure that no increased risk of pollution to the water source occurs. DoW uses a "risk minimization" approach for P2 areas.
3. *Priority 3 (P3) Areas* are declared where water supply sources need to coexist with other land uses, such as residential and commercial, and DoW considers that risk management is required to avoid pollution of the water sources. DoW uses a "risk management" approach for P3 areas.

Recreational use of water catchments in WA has been restricted largely to P2 and P3 catchments, dams supplying irrigation water to agricultural areas, and a small number of water catchments yet to be proclaimed as public drinking water source areas (PDWSAs). The security of recreational access to irrigation dams is not mandated and could be withdrawn at any time by water agencies.

Water Corporation is responsible for the capture and supply of clean drinking water to WA urban areas through a reticulated system. Supply comes from water captured in a series of public drinking-water dams and is also sourced from groundwater reservoirs. Water Corporation's management of drinking-water catchment areas is directly governed by the policy of DoW, and it is the agency responsible for enforcement of restricted access to catchments.

Several other agencies and organizations, such as the Department of Sport and Recreation, Department of Health, and Tourism WA, do not have direct land or water management jurisdiction yet nevertheless also seek to influence policy regarding recreational access to WA drinking-water catchments and lobby the government based on their own legislated mandates. The Department of Health is concerned with minimizing disease risk and encouraging healthy lifestyles. The Department of Sport and Recreation is concerned with promoting physical activity and providing opportunities for outdoor recreation. Tourism WA focuses on marketing opportunities for tourism-related activity and associated economic benefits to WA.

Legislation and Policy

The legislative and policy context associated with water catchments strongly influences how they are managed. A range of WA state legislation, policies, and strategies relate to the management of PDWSAs and human activity in those areas. Fifteen statutes and policies have been identified that govern the management of catchments in terms of water supply and recreation in the southwest of WA (Hughes et al. 2008). The various government agencies are mandated by different acts, bylaws, and regulations related to the management of water catchments. In some areas, responsibilities overlap among management agencies as determined by the relevant laws. When land and water management responsibilities overlap, the primacy of legislation governing management roles can be uncertain and need to be clearly defined. This is important for avoiding clashes of management approaches or duplication of activity. The general rule for establishing primacy of legislation is that later provisions take primacy over earlier ones (Gifford 1990). However, laws and related sections can be amended frequently, causing confusion over which legislation is most recent and therefore has primacy.

Importantly, water catchments in the Darling Range mainly fall within protected areas (Ingram 2009). WA protected areas are managed by DEC according to the Conservation and Land Management Act 1984 WA (CALM Act) (Ingram 2009). DEC manages these public lands for many diverse values, as determined by the CALM Act, including nature conservation, recreation, water catchment protection, and timber production (CCWA 2003). The catchment areas include mining tenements and private land, the latter of which is commonly used for agriculture but also includes urban development. Mining tenements in the Darling Range are primarily associated with bauxite mining using strip-mining methods, where a network of relatively small, open-cut pits connected by access and haul roads is created within a given area of forested catchment. These various land uses can have significant impacts on water quality in the catchment.

In drinking-water catchments, the CALM Act can overlap with two other main acts relating to WA drinking-water source protection: the Country Areas Water Supply Act 1947 (CAWS Act) and the Metropolitan Water Supply, Sewerage and Drainage Act 1909 (MWSSD Act), together referred to as the water supply acts. These laws grant extensive powers to DoW to proclaim PDWSAs and include elements that run counter to the CALM Act. The water supply acts also provide DoW with power to implement control measures to safeguard the water resources in these catchments. This includes controlling the type and extent of recreation in P2 and P3 water catchments through the preparation of Drinking Water Source Protection Plans and the associated gazettal[1] of these water catchments, and excluding recreational access from P1 areas, to achieve water quality objectives (Hughes et al. 2008).

DoWs policy on recreational use of water source catchments on Crown land (Policy 13) defines recreation as "a wide range of leisure, pastime or entertainment pursuits, including bushwalking, orienteering, swimming, boating, fishing, camping, horse-riding and four-wheel driving" and also includes "group outings and commercial activities such as guided tours and car rallies." The policy states that the final approval function for all recreational access lies with DEC. It also states that access is governed by the two acts under the portfolio of the minister for water resources. Policy 13 makes clear that the management and approval functions lie with DEC, and that DEC is required to take into consideration the water protection acts in granting approval for recreational access and carrying out its management functions in catchment areas. DEC, in turn, has extensive powers to make policy statements and statewide plans to guide it in doing so.[2] In practice, however, the situation is less clear in land areas where the divergent mandates of the water supply acts and CALM Act overlap, legislative primacy is uncertain, and there is little or no allowance for collaboration among management agencies.

Clash of Responsibilities

When clashing legislation overlaps, and no provisions that define a course of action are apparent, establishing which act has primacy is particularly important. This is evident in the difficulties presented by the independent planning and management processes undertaken by DEC and DoW for the same catchment areas. DEC undertakes its catchment planning through the Forest Management Plan and various national park plans. DoW creates its water source protection plans through separate and distinct processes based on the CAWS and MWSSD Acts. No formal processes are in place by which both DoW and DEC could jointly assess proposed management measures for the same catchment area. Subsequently, as water is currently valued more as a consumptive resource, the control measures devised by DoW under the powers granted to it by the CAWS and MWSSD Acts have come to be accepted without adequate debate regarding their practicalities in terms of implementation and encroachment on recreation and tourism uses. These planning processes need to be better coordinated and integrated for effective management and clarity of responsibility on the ground. This could be facilitated through the Forest Management Plan, which has a recreational planning in water catchments framework. Integration of water quality management and recreation management planning is possible in catchments, as evidenced by the Victorian example discussed below.

To integrate water protection and recreation planning for catchments, adoption of a risk management approach by DoW and Water Corporation is necessary. Risk management is based on managed access of recreation in and around water catchments and is a widely accepted practice globally (Pigram 2006). Such a regime could foster a supportive recreational community that would be more open to a positive communication and education campaign aimed at responsible use in defined areas, at given times, or both (i.e., with spatial and temporal constraints). Work in the United Kingdom and the United States provides considerable evidence to support the view that properly planned and managed recreation in water catchments can be acceptable on public health grounds (Hughes et al. 2008).

However, this would require a fundamental shift in DoW's stance on water protection away from a risk avoidance management culture. This could prove difficult, as the links between varying forms of recreation and water quality in drinking-water catchments are unclear. Hammitt and Cole (1998) point out that recreation impacts on water quality are generally site-specific and cannot be generalized to all circumstances. They also note a number of contradictory studies published in relation to recreation and water quality. Because of the site-specific character of impacts, monitoring is complex, time-consuming, and expensive, rendering studies of this nature impractical for limited government budgets (DoH 2007). Consequently, catchment management action relies on expert opinion and modeling based on scientific inference, leading to considerable margins of error in assessing risk.

As a means of assessing risks to water quality from recreation, the Advisory Committee on Purity of Water (1977) compiled a list of recreational activities along with potential risks and suggested management actions, based on observations in WA catchments (see Table 9.1).

DoW and Water Corporation currently classify all recreation as a single type of activity that poses an unacceptable risk to water quality. However, the 1977 expert review found that certain "passive activities" posed a low risk, while other activities were high-risk. This general understanding of risks from recreation, combined with the fact that drinking-water contamination events are rarely related to recreational use of catchments, could form the foundation for integration of catchment management between DoW and DEC, which is currently lacking in WA (Hughes et al. 2008). That is, a shift in water catchment management thinking toward recognition that recreation consists of a range of activities along a spectrum of risk to water quality is essential for better integration of catchment management. This approach could allow low-risk recreational access (such as walking and picnicking) while restricting or prohibiting high-risk activities (such as powerboats, swimming, and fishing).

Table 9.1 *Recreational activities and assumed risk to water quality in catchments*

Recreational activity	Potential risk/management action suggested
Horse riding	Excretion of coliform bacteria and salmonella poses a contamination risk; should not be allowed in or near water bodies and running streams
Four-wheel-drive vehicles	Increased turbidity in streams; should not be permitted off-road in watersheds
Motorcycles	Nuisance value to other catchment users, damage to soft terrain; certain portions of watersheds could be made available to organized trail-riding events, contingent on monitoring
Picnicking	Low risk, given provision of suitable toilet facilities
Fishing	Use of bait and "unfortunate personal habits" pose a considerable health risk
Canoeing	None noted, as this was a "relatively new custom"
Swimming	"Physical (and possibly psychological) effects" are undesirable; should remain prohibited in reservoirs
Powerboats	Erosion of bank and shore areas, disturbance of fish and bird breeding areas, oil leakage and exhaust fumes; should not be permitted on reservoirs

Source: Adapted from Advisory Committee on Purity of Water (1977, 8–14)

Catchment Governance in Western Australia

As evidenced by the legislation and resultant planning processes, the governance of southwest WA water is dominated by government. The corporate, top-down approach to water governance in WA, combined with a focus on consumptive values, means that water management agencies and policymakers have been under no compulsion to consider recreation as a water user. As Pigram (2006) notes, "It is readily apparent that most water management authorities regard recreation as ancillary." As a result, the need for research to understand the sociocultural values of instream uses of water has been largely ignored. Consequently, insufficient attention has been given to such uses in water resource decisionmaking. This lack of recognition and research base has inhibited the potential for a shift toward multiple-use management of WA water catchments.

As an example, the draft Southwest Regional Water Plan (DoW 2008) failed to consider the nonconsumptive uses of water, such as recreation and tourism, as legitimate and appropriate water uses. The management of water-dependent features, such as reservoirs, dams, lakes, wetlands, rivers, streams, and catchments, for cultural, heritage, recreational, and tourism uses has not been adequately recognized, and the water needs of these sectors have not been not planned for and integrated with other water uses. Consequently, the recreation and tourism industries have been unable to influence important water-planning processes that either directly or indirectly affect recreation and tourism opportunities. Managing recreation in water catchments in WA will require a cooperative governance approach involving a range of government agencies, interested stakeholders, nongovernmental organizations, and the community.

LESSONS FROM VICTORIA

In Victoria, by contrast, drinking-water catchments are managed under a varied system of restricted and open public access (Hughes et al. 2008). Forested catchments in the Yarra Ranges provide 90% of Melbourne's drinking water. Water from the Yarra Ranges is pumped through a supply system to three retail water companies for distribution throughout Melbourne (Melbourne Water 2009). Drinking water is supplied to areas outside metropolitan Melbourne through 15 regional water authorities. The drinking-water suppliers are regulated by the Department of Human Services, Department of Sustainability and Environment, and Office of the Regulator-General (Hughes et al. 2008).

Parks Victoria manages national parks in the state according to the National Parks Act 1975. As a primary catchment for Melbourne's drinking-water supply, almost 85% of the Yarra Ranges National Park is a designated water supply catchment area. This area is legislated under the National Parks Act 1975 to protect water catchment and water resource values. Based on this designation, recreation is restricted and managed jointly by Parks Victoria and Melbourne Water in accordance with a catchment management agreement. The National Parks Act includes a provision that applies only to the areas that are part of the drinking-water catchment for Melbourne (Ingram 2009). The act also provides powers to ensure that the natural and other features of the national park are protected and public use for tourism and recreation provided for. In relation to this, a catchment management agreement was signed in 1995 between the director of national parks and Melbourne Water. The agreement provides the basis for cooperative management of the drinking-water catchment and determines the respective management responsibilities of Parks Victoria and Melbourne Water (Parks Victoria 2002).

This partnership governance model is based on the notion that it is more efficient to first protect water quality to ensure an initial high standard rather than to treat it later to reach the required quality standards (Melbourne Water 2009). As a result, public access is prohibited in many areas in line with this approach. This is also a result of a historical management practice, where 157,000 hectares of forest has been closed to the public for over a century (Ingram 2009). This governance approach was modified and strengthened with the National Parks Act 1975, which provided clear powers to manage Melbourne's water catchments for the primary objective of water quality first and foremost, and then recreational access and use only where it is considered not to affect water quality. The fact that the partnership between Parks Victoria and Melbourne Water is provided for through legislation appears to strengthen this governance approach, as the two entities rely on each other to achieve their collective visions of protection: that is, multiple-use management that delivers water quality for both drinking purposes and ecosystem preservation.

CONCLUSIONS

Achieving a balance among recreation, tourism, water quality and quantity, and other catchment values requires careful consideration of public health, social, economic, and environmental factors. However, recreation, tourism, and other social considerations are given little credence in the water-planning processes in WA. New and inclusive measures are required for planning and management of catchments to fully consider all these elements.

Integrating the planning for water protection, recreation, and tourism in water catchments is complex. In addition to overlapping legislation and responsibilities, as well as lack of clarity regarding legislative primacy, the historical, social, and political contexts play significant roles in catchment management and governance. Furthermore, different management organizations may have what could be described as clashing cultures. For example, in WA, both DoW and Water Corporation view all recreation and tourism as a uniform contamination threat, whereas DEC perceives varying levels of risk that could be managed in a range of activities. Catchment management needs to move away from a solely threat-based approach and adopt values-based management practices.

The current threat-based approach in WA is based on presumption rather than solid evidence. This has resulted in policy and management focused on public exclusion at all costs to ensure provision of cheap, clean drinking water. Those players responsible for protection of water quality for drinking in WA are driven by a view that recreation is a single, uniform activity posing a uniform threat. However, it is clear that different types of public access present varying levels of risk to water quality. Recognition of recreation as a diverse range of activities with varying threat levels could provide a basis for better-integrated catchment management in WA. Integration of PDWSA catchment management with protected-area management for conservation and recreation would work to recognize the needs and resulting benefits to the fourth "player," the groups and individuals using catchments for tourism and recreation.

Recreational users accessing water catchments are driven by a range of wants and needs. For example, most seek good quality experiences, adequate on-site facilities, and minimal conflict among recreation types. Some forms of recreation require specific facilities, such as dedicated tracks or vehicular access. Some groups focus more on the aesthetic and ecological quality of the catchments they access (Ingram and Hughes 2009). Nevertheless, all are members of the broader community who also expect to have access to clean, safe drinking water in their homes on a daily basis. Integrated, values-based management could incorporate recognition of water catchments as both sources of clean drinking water and areas with social, recreational, and tourism value.

NOTES

1. To "gazette" means to gain government approval and recognition in law. A gazetted area is one that is approved by law for a particular use and managed according to specific legislation, regulations, and policy, and therefore falls under the responsibility of specific government departments.

2. See, for example, CALM Act sections 19(1)(c), 33(dd), and 55.

REFERENCES

ACPW (Advisory Committee on Purity of Water). 1977. *A Study of Catchments and Recreation in Western Australia*. Perth: Working Group on Catchments and Recreation.

CCWA (Conservation Commission of Western Australia). 2003. *Forest Management Plan 2004–2013*. Perth: Conservation Commission of Western Australia.

DCLM (Department of Conservation and Land Management). 1992. *Management Strategies for the South-West Forests of Western Australia: A Review*. Perth: Department of Conservation and Land Management.

———. 2002. *Draft Forest Management Plan*. Perth: Department of Conservation and Land Management.

Dearden, P., M. Bennett, and J. Johnston. 2005. Trends in global protected area governance, 1992-2002. *Environmental Management* 36 (1): 89–100.

DEC (Department of Environment and Conservation). 2004. *Land Use Compatibility in Public Drinking Water Source Areas. Water Quality Protection Note*. Perth: Department of Conservation and Land Management.

DoH (Department of Health). 2007. *Recreational Access to Drinking Water Catchments*. Perth: Department of Health.

DoW (Department of Water). 2008. *Western Australia's Achievements in Implementing the National Water Initiative*. Perth. http://breeze.water.wa.gov.au (accessed July 2010).

Feilman Planning Consultants. 1987. *Recreational Opportunities of Rivers and Wetlands in the Perth to Bunbury Region. Wetlands Usage Report*. Volume 1. Perth: Water Authority of Western Australia.

Forests Department Northern Region. 1983. *Forest Recreation Framework Plan*. Perth: Western Australia Forests Department.

Gifford, D. 1990. *Statutory Interpretation*. Holmes Beach, FL: Gaunt.

Hammitt, W., and Cole D., 1998. *Wildland Recreation: Ecology and Management*. New York: John Wiley and Sons.

HCN/RMNO (Health Council of the Netherlands and Netherlands Advisory Council for Research on Spatial Planning, Nature and the Environment). 2004. *Nature and Health: The Influence of Nature on Social, Psychological and Physical Well-Being*. The Hague: Health Council of the Netherlands and RMNO.

Herath, G. 2002. The economics and politics of wilderness conservation in Australia. *Society & Natural Resources* 15 (2): 147–59.

Hughes, M., Zulfa M., and Carlsen J., 2008. *A Review of Recreation in Public Drinking Water Catchment Areas in the Southwest Region of Western Australia*. Perth: Curtin Sustainable Tourism Centre, Curtin University.

Ingram, C. 2009. *Governance Options for Managing Sport and Recreation Access to Water Sources and Their Catchments of the Southern Darling Range, Western Australia*. Perth: Resolve Global Pty. Ltd.

Ingram, C., and Hughes M., 2009. *Where People Play: Recreation in the Southern Darling Range, South Western Australia*. Perth: Resolve Global Pty. Ltd.

Maller, C., Townsend M., St Leger L., Henderson-Wilson C., Pryor A., Prosser L., and Moore M., 2008. *Healthy Parks, Healthy People: The Health Benefits of Contact with Nature in a Park Context*. Melbourne: Deakin University.

Martinick & Associates. 1991. *A Review of the Water Based Recreation in Western Australia*. Perth: Ministry of Sport and Recreation and Western Australian Water Resources Council.

Melbourne Water. 2009. *Water Catchments: Supply Storage Areas*. www.melbournewater.com.au (accessed May 28, 2009).

Muench, R. 2001. *Southern Darling Range: Regional Recreation Study*. Perth: Department of Conservation and Land Management, Water Corporation and Water and Rivers Commission.

Murdoch University. 1985. *Waroona and Logue Brook Reservoirs Environment and Recreation Study*. Perth: Murdoch University.

Natural England Board. 2007. *Draft Health Policy Position Statement*. London: Natural England Board.

Newman, J., Barnes M., Sullivan H., and Knops A., 2004. Public participation and collaborative governance. *Journal of Social Policy* 33: 203–23.

Page, B., and Kaika M., 2003. The EU water framework directive, Part 2: Policy innovation and the shifting choreography of governance. *European Environment* 13: 328–43.

Pahl-Wostl, C., Craps M., Dewulf A., Mostert E., Tabara D., and Taillieu T., 2007. Social learning and water resource management. *Ecology and Society* 12 (2): 5.

Parks Victoria. 2002. *Yarra Ranges Management Plan*. Melbourne: Government of Victoria.

Patterson, J. 1977. *Alternative Recreational Use Policies for System 6 Reservoirs*. Perth: University of Western Australia.

Pigram, J. 2006. *Australia's Water Resources: From Use to Management*. Victoria: CSIRO Publishing.

Sharp, J. 2005. Healthy parks, healthy people. *Landscope* 4: 27–31.

Stephenson, G., and Hepburn J.A., 1955. *Plan for the Metropolitan Region Perth and Fremantle Western Australia. Report*. Perth: Government Printing Office.

SWALSC (South West Aboriginal Land and Sea Council). no date. *History of the Noongar*. www.noongar.org.au (accessed August 18, 2010).

Thorpe, C. 2006. *Lost Rivers*. Harvey, Western Australia, Community Forum on Logue Brook Dam, July 22, 2006.

UN (United Nations). 2009. *What Is Good Governance?* New York: United Nations Economic and Social Commission for Asia and the Pacific.

Weller, R. 2009. *Boomtown* 2050: *Scenarios for a Rapidly Growing City.* Perth, Western Australia: UWA Press.

WRC (Water and Rivers Commission). 2003. *Statewide Policy No. 13: Policy and Guidelines for Recreation within Public Drinking Water Source Areas on Crown Land.* East Perth: Water and Rivers Commission.

Yates, C., A. McNeill, J. Elith, and G. Midgley. 2010. Assessing the impacts of climate change and land transformation on banksia in the South West Australian Floristic Region. *Diversity and Distributions* 16: 187–201.

CHAPTER 10

Cases in Policy Suasion and Influence: The Boating Industry

Sue O'Keefe and Glen Jones

C onventional economic theory would hold that, given the existence of water markets, competing economic and other interests can be encapsulated in the individual decisions to buy or sell water rights. As was noted in the introduction to this book, this is relatively clear-cut in the case of agriculture, where the relationship between water as an input and an output is well established and understood. In the case of recreation and tourism, however, the matter becomes more complex. Recreational boating, for example, takes place outside of the market space, and therefore the value of water as an input is not revealed. Despite the various attempts to ascribe a value to water for tourism and recreation, in practice Australia has to date been left without a market price and therefore must rely on political suasion and lobbying to achieve desired policy outcomes. In some instances, the interests of boating enthusiasts might coincide with those of irrigators, such as where boating is dependent on reservoirs maintained for the purposes of irrigation. In contrast, some boating activities are negatively affected by excessive extractions from an already overstretched river system such as in the Murray-Darling. Similar convoluted relationships exist between boating and environmental interests. The existence of these relationships raises the possibility of the formation of alliances to influence policy outcomes.

In this chapter, we present two case studies that serve to illustrate the manner in which the Boating Industry Association of South Australia (BIASA) and the United States National Marine Manufacturers Association (NMMA) have set about harnessing important complementarities with other users to provide the suasion to influence the formulation of water policy and legislation. These case studies exemplify alternate approaches to gaining influence at the policy table and provide practical scenarios which are built on in Chapter 11. This relatively brief chapter provides a practical rather than theoretical view and draws heavily on personal experience. It begins with a brief description of some of the hydrological and environmental realities in South Australia, then examines each of the case studies in detail.

HYDROLOGICAL AND ENVIRONMENTAL ISSUES OF SOUTH AUSTRALIA

The problems associated with a dwindling and overallocated freshwater resource are arguably no more apparent than in South Australia. In this state jurisdiction can be seen in sharp relief the serious impacts of competing extractive and nonextractive users who vie for their share of this resource. Although agriculture has historically exerted the most predominant claim on the resource, there are also undeniable environmental imperatives, particularly in the Coorong. The problem is further confused by the importance of the tourism and recreational sector along this stretch of the Murray, and it is in this context that the Boating Industry Association of South Australia has exerted significant influence.

South Australia is home to more than 1.6 million people, with the majority located in the capital Adelaide, along the southeast coast or in close proximity to the River Murray. Its main industries include mining, agriculture, and manufacturing. Adelaide and much of South Australia rely on the River Murray for their water supplies. In a typical year, Adelaide draws about half of its water needs from the River Murray, and this increases to 90% in a dry year.

The Coorong, Lower Lakes, and Murray Mouth (CLLMM) is a 140,000-hectare complex of lakes, streams, lagoons, and wetlands at the end of the Murray-Darling river system. The CLLMM was listed as a "wetland of international importance" under the Ramsar Convention in 1985. The system relies mainly on water from the River Murray, which fills the Lower Lakes (Lake Alexandrina and Lake Albert) before discharging to the Southern Ocean, via the Coorong and Murray Mouth.

The volume of water supplied to the CLLMM has fallen over time as a result of upstream development. In 1940, tidal barrages were constructed near the Murray Mouth to limit seawater intrusion. Kingsford et al. (2009) note that the lakes were generally freshwater prior to barrage construction, with some minor saltwater incursions.

Inflows to the CLLMM were lower than usual in the 1990s and have since worsened, with record low inflows between 2006 and 2009. As a result, the water level in the Lower Lakes was below sea level between 2007 and 2009, and the Coorong has become increasingly saline. Average daily salinity at Tauwitchere Barrage roughly doubled between 2005 and 2008 (DWLBC 2010). The CLLMM has experienced a decline in species diversity and abundance. In the Lower Lakes, there has been a change toward estuarine and marine animals, which are better suited to the altered conditions. Conditions have also become more favorable for serpulidae, a marine tubeworm that attaches to hard surfaces, killing freshwater mussels and turtles. The number of waterbirds was estimated to have fallen by around 50% between 2007 and 2008 (Kingsford et al. 2009).

Box 10.1. *Challenges of Overextraction and Competition*

Put simply, reduced water levels mean that if your boat or ferry needs a meter of water in which to float, it won't float when the water levels fall below one meter. When the water level falls, you can't get up to the bank to load provisions or passengers; you can't get into or, if you're unlucky, out of the marina; you can't get to the sewage pump out point or the petrol pump on the bank; the water at the deep end of the ramp isn't sufficient to float off the trailer; and the ferry can't load the vehicles from the road. Snags, rocks, and other hazards that have never been seen before now poke through the surface or, worse, lie just out of sight below the surface.

Another concern is the exposure of acid sulfate soils around the lakes' edges. Soils on the lakebeds naturally contain iron sulfides, which produce sulfuric acid when exposed to air. Acid sulfate soils have been exposed as water levels have fallen. When these soils are covered with water again, large amounts of sulfuric acid could be released into the lakes. This could contribute to the mobilization of metals, decrease in water oxygen levels, and production of noxious gases (MDBA 2010). Kingsford et al. argue that the Lower Lakes have a natural alkalinity that should buffer the effects of acidification; however, this capacity is limited. They also note that acidification is sometimes a natural process: "it also occurs episodically and sometimes seasonally under natural conditions and triggers fish kills that do not necessarily lead to long-term ecosystem damage under natural conditions" (2009, 46).

For the boating industry of South Australia, which relies on the health of the river, the challenges of overextraction and competition among users pose a very real threat, as described by one boating enthusiast in Box 10.1.

CASE STUDY: THE BOATING INDUSTRY ASSOCIATION OF SOUTH AUSTRALIA (BIASA)[1]

Recreational boating is a nontrivial economic activity, with some estimates of its contribution to the economies of the Murray-Darling Basin exceeding A\$6 billion.[2] The health of waterways is a key requirement for this activity. In particular, in South Australia, the River Murray is home to a number of boating-related recreational activities such as fishing, water-skiing, cruising and touring, hunting, houseboating, swimming, bird-watching, and a host of others.

Boat types range from very small, one-person tinnies (open aluminum boats) and inflatables to large craft that accommodate more than 100 overnight guests who are served by dozens of onboard staff members. More than 1,000 cruising craft over 8 meters in length provide galley and overnight accommodations in the waters of the Lower Murray and Lower Lakes in South Australia. The great majority of these craft are serviced and provisioned within the local communities. Thousands of holiday homes and shacks, together with their attendant recreational boats, line the banks of the river and lakes. Major marinas containing many hundreds of recreational craft are also found along this last reach and within the Lower Lakes. Twenty communities located within five local government areas depend, to a large extent, on the ongoing success of the recreational boating sector for their economic viability.

The interests of public boating are overseen by the Boating Industry Association of South Australia (BIASA). BIASA is the main industry body in South Australia for recreational and light commercial boating; it is one of six stakeholders of the Australian Marine Industries Federation (AMIF), the main industry body in Australia as a whole for recreational and light commercial boating. In addition, the AMIF is involved in the operation of the Australian International Marine Group, which choreographs the efforts to increase national marine exports (A$1 billion annually), and the Marina Industries Association of Australia, which is the national voice for the country's marina and slipway operations, with approximately 1,000 operators.

The AMIF represents the interests of more than 2,000 businesses and 5 million Australians. It has a diverse membership, including individual business interests such as boat builders, naval architects, charter operators, hire operators, historic wooden boat builders, and marina operators. It also has organizational members such as industry development organizations, industry training bodies, and insurance and finance organizations.

Not surprisingly, the interests of the boating industry at times coalesce, and at other times conflict, with other resource claimants, such as agriculture and environmental interests. One reason for the tourism and recreation sector's lack of influence is that it is often viewed as a bunch of hedonists with too much spare time on their hands. In the case of the Coorong, Lower Lakes, and Murray Mouth described above, however, the interests of environmentalists and boaters merged as the environmental challenges of the River Murray were also accompanied by nontrivial impacts on the recreational boating industry. Within BIASA as early as May 2000, fears were raised about lowering water levels in this region. As a result, the ministers for tourism and transport commissioned BIASA to conduct a two-week field exercise on the 570 kilometers of the River Murray from Lake Alexandrina upstream to the South Australian and New South Wales border. Participants in this exercise included representatives of local government, scientists, members of parliament, industry group representatives, and government bureaucrats.

The purpose of this exercise from the state government's perspective was to improve the safety of navigation along the river for tourists. BIASA, however, seized on the opportunity to develop key relationships that would prove crucial in raising awareness of its own interests. In preparation for the venture, BIASA invited state agencies and local government bodies—including the Environmental Protection Agency; the Departments of Environment and Heritage, Water, Land

and Biodiversity Conservation, and Transport; and the SA Tourism Commission—to list their main concerns in relation to the River Murray. Not surprisingly, these concerns generally related to the health of the environment and included the condition of national parks, stock fencing, hazard signs, and the location of snags and weeds. The exercise also involved plotting the placement of navigation markers, recording interpretive signage, and with consideration of the fragility of banks and environment, identifying locations to be defined as safe moorings. In addition, BIASA reviewed other issues, such as the location and condition of waste disposal and pumping stations, services provided by lockkeepers, and the ease of access and maneuverability within locks. It also identified locations of privately owned vessels and reported on environmental problems.

BIASA was so successful that it was commissioned to carry out similar tasks in ensuing years, with the area of survey being extended to include Lakes Alexandrina and Albert and the Coorong. This type of consulting work has now expanded to account for 40% of BIASA's income. The most tangible product from the exercise was *South Australia's Waters: An Atlas and Guide*, which was produced and released in 2005. Five thousand copies were sold, many destined for government purposes, and a second edition was released in 2008. The wide use of this atlas within government agencies reinforced the standing of BIASA and gave it an improved level of access to federal, state, and local government bodies.

The cooperative work undertaken by BIASA has resulted in improved understanding among the various sectors. This type of coproduction of knowledge is considered in depth in Chapter 11, but several important relationships forged from this exercise are outlined here. First, BIASA has developed and maintained close relationships with a number of government departments, including Aboriginal Affairs and Reconciliation, Education, Consumer Affairs, and Treasury. each of which shares some common interests in river health with BIASA. Second, BIASA was invited to provide representatives for the Murray-Darling Basin Commission's Community Reference Panel, which had consisted mainly of agricultural interests. This led to increased understanding between these two groups whose interests were previously seen as being at odds. One result of this newly forged relationship was BIASA's development of a recreational boating industry code of practice, which took into consideration the interests of both agriculturists and environmentalists, and included initiatives such as regulations on the installment of graywater treatment systems on recreational boats. Third, BIASA established reference groups based at opposing ends of the River Murray, comprising members of local and state government agencies with an interest in drought matters and local members of state parliament. These groups met formally each month to discuss issues of concern and how they could be resolved. Fourth, the work of BIASA has precipitated a number of appointments to government committees and working parties where a degree of influence can now be wielded. BIASA now participates within water policymaking bodies in all three levels of government, enabling the tourism and recreational sector to be considered in future water policy decisions.

Although the gradual increase in policy influence of BIASA may have been simply opportunistic, the following example from the United States contains a very strategic and planned approach to achieving policy change. It involves a multipronged approach that included efforts on the legislative, judicial, regulatory, and media fronts and a proactive campaign designed to achieve its objectives.

CASE STUDY: THE NATIONAL MARINE MANUFACTURERS ASSOCIATION (NMMA)[3]

In September 2006, a US District Court ruling nullified US Environmental Protection Agency (EPA) regulation 40 CFR 122.3(a) under the Clean Water Act (CWA), which exempted effluent discharges incidental to the normal operation of vessels, including recreational boats, from the National Pollutant Discharge Elimination System. The ruling resulted from a lawsuit brought by environmentalists and states to halt the introduction in US waters of invasive species through commercial ballast water. Although the court never contemplated incidental water discharges, as the case centered on commercial ship ballast water, the effect of its decision vacated the exemption for these nonpolluting discharges. The discharges included under the original, longstanding federal exemption were engine cooling water, graywater, uncontaminated bilgewater, and deck runoff. The US District Court ruled that EPA lacked the authority under the CWA when it originally established this exemption, although EPA's regulatory exemption had been in place for more than 35 years without challenge.

Under the court order, EPA was forced to initiate a massive new federal regulatory scheme and establish permit requirements for the normal discharge of every recreational boat by September 2008. EPA began in haste with the development of this regulation through a formal rulemaking process to meet the court deadline. The agency and states were set to create a new, onerous, and unprecedented enforcement and permit regime for every recreational boat owner in the United States, in addition to all commercial vessels, for all incidental effluent discharges. Additionally, boaters would now be legally exposed to individual citizen suits under the CWA.

The NMMA's Reaction, Strategic Planning, and Financing

Recognizing the severity of this court decision, the National Marine Manufacturers Association (NMMA) immediately began a strategic-planning process and developed a long-term plan to reinstate the original federal exemption prior to the September 2008 deadline. This plan entailed engaging Congress to pass legislation reinstating the exemption for incidental water discharges for recreational vessels; engaging EPA throughout the formal rulemaking process to mitigate the negative impacts of its new permitting regime should the legislative effort fail; participating in an appeal of the US District Court decision; building an all-industry and third-party coalition to support the legislative advocacy campaign; and engaging the media to leverage public awareness to support the advocacy and grassroots campaign.

The NMMA plan was multipronged and concurrent by design, with the purpose of achieving success in the US Congress and the judicial and executive branches. Should any one effort fail, the others would serve as a backstop. NMMA sought direction and funding for its plan from its board of directors. Once the plan was approved and funding was obtained, the association began what would be a two-year process to undo the District Court decision. As the national trade association for the marine industry, NMMA provided the main financing and leadership on the initiative, although coalition support was crucial to the effort.

Coalition Development

The most crucial aspect of achieving legislative success was clearly the development of a significant coalition, built by NMMA from scratch. To that end, NMMA sought the assistance of BoatU.S., which had 650,000 members, and these two organizations established the Boat Blue coalition. Ultimately, Boat Blue comprised more than 60 organizations throughout the United States, including conservation and angling groups, state marine trade groups, boating organizations, and others. NMMA managed all communication and messaging to the coalition, financed and created a website specific to the issue with advocacy tools, and marshaled enormous grassroots support. After NMMA's campaign resulted in the introduction of the Clean Boating Act in Congress, more than 250,000 messages were sent to Congress urging the act's expeditious passage. This grassroots activism was essential to eventual passage of the bill and was possible because of coherent messaging, clear leadership, and the willingness of all coalition members to work jointly in mobilizing their members and contacts.

Media Campaign

To support its overall initiative, NMMA aggressively engaged the media to leverage public support. NMMA established a recognizable brand for the campaign in Boat Blue and proactively marketed its initiative. The effort led to millions of impressions[4] and major favorable stories over a two-year period in top media outlets throughout the nation. The media and messaging efforts were crucial to the overall success of NMMA's endeavor.

Legislative Effort

The chief concern for the industry was to reinstate the exemption for incidental discharges through an act of Congress. Initial efforts resulted in the introduction of the Recreational Boating Act in the House of Representatives. A Senate version of this legislation was later introduced, but this legislation lacked the support of key Senate leaders and was strongly opposed by the environmental community. As a result, NMMA negotiated with environmental groups and Senate committee leaders on more viable legislative language. The result was that Congress passed the Clean Boating Act on July 22, 2010, and President Obama signed it into law on July 30, days before the new EPA permitting regime would have gone into effect.

Regulatory Effort

Concurrent with the legislative effort, NMMA sought to influence the outcome of the ongoing EPA regulatory development of the court-ordered permitting scheme in the event that legislation ultimately failed to pass. To accomplish this, NMMA built a lasting relationship with the EPA team charged with developing the regulation, provided extensive formal written comment for the record, and led a joint effort with the Boat Owners Association of the United States (BoatU.S.) and other industry partners, which led to thousands of individual boaters submitting public comment to the docket in support of the NMMA position. These comments were entered into the public record through NMMA's online advocacy toolkit, which made it easy for individual boaters to comment. The EPA administrator and other agency officials also spoke at NMMA's Washington, DC, conference, and NMMA led an educational trip for EPA staff to the Miami Boat Show to familiarize them with recreational vessels and the industry.

Judicial Effort

Because the original litigation, *Northwest Environmental Advocates, et al. v. U.S. EPA*, centered on commercial vessel ballast water regulation, NMMA was not a direct party in the suit. However, NMMA, using outside and in-house counsel, filed an extensive *amicus curiae* brief with the appellate court recommending that the initial judgment be narrowed. Other affected parties, such as commercial ship operators, also entered the litigation process on appeal. The decision of the appellate court was not completed until after the Clean Boating Act of 2008 was signed into law.

CONCLUSIONS

The examples provided in the two cases above serve to highlight the importance of a collaborative approach to achieving policy outcomes that are valued by the stakeholders. Importantly, and particularly in the case of BIASA, the role played by those with on-ground knowledge is crucial. The role of knowledge and the way in which it is framed are given further attention in Chapter 11. In that chapter, drawing on the case studies presented in this and the previous two chapters, Ronlyn Duncan develops a theoretical framework that is seen as potentially beneficial for the tourism and recreation sector.

NOTES

1. This case study draws heavily on information provided by Glen Jones, general manager of the Boating Industry Association of South Australia.

2. A\$1 = US\$0.9978 as of January 2011.

3. The authors thank Thom Dammrich, president and CEO of the National Marine Manufacturers Association, based in Chicago, for providing information for this case study.

4. Impressions are a metric referring to the number of people who heard or read about the issue in the media, such as in a newspaper or journal article, on the radio or television, or on the Internet.

REFERENCES

DWLBC (South Australian Department of Water, Land and Biodiversity Conservation). 2010. *Natural Resources Information Management System*, accessed September 29, 2010, from http://e-nrims.dwlbc.sa.gov.au.

Kingsford, R., P.G. Fairweather, M.C. Geddes, R.E. Lester, J. Sammut, and K.F. Walker. 2009. *Engineering a Crisis in a Ramsar Wetland: The Coorong, Lower Lakes and Murray Mouth, Australia.* Sydney: Wetlands and Rivers Centre, University of New South Wales.

MDBA (Murray-Darling Basin Authority). 2010. *Acid Sulfate Soils Field Guide*, accessed September 29, 2010, from www.mdba.gov.au.

CHAPTER 11

Science, Policy, and Knowledge: Is There a Better Way for the Tourism and Recreation Sector?

Ronlyn Duncan

This chapter addresses the question of whether there is a better way for the tourism and recreation sector to harness the knowledge it needs to seek and secure adjustments to existing water policy. This is a crucial question for a sector that has a limited knowledge base from which to promote or defend its claims for access to water resources (Pigram 2006). For example, Chapter 9, in its analysis of recreational access to urban water supply catchments, argued that the significant value of recreation—the social, cultural, and economic benefits of people being active in life and communing in the outdoors in and near water—is not fully understood, nor is it adequately reflected in water policy in Australia. Recreationists have increasingly been denied access to areas where they previously were allowed, and areas that could be managed for public access are not. Although exceptions do exist, in Australia the rule appears to be prohibition for the protection of water quality and public health. Chapter 9 identified clashes of jurisdiction and legislative objectives, as well as a lack of research on the impacts of recreation on water quality, as entrenching the denial of access to public land for recreation.

Adding to the challenge is the diffuse nature of the sector and the complexity of its water needs (ABS 2009; Pigram 2006). For example, as discussed in Chapter 5, most sector businesses do not have property rights to the water inputs of their products. This means that enterprises dependent on tourism and recreation trade have limited capacity to manage declines in water availability or its reallocation elsewhere. Importantly, opportunities for recalibrating existing water policy settings often are not immediately apparent. Possibilities might arise from direct engagement with property rights holders, or marginal adjustments to property rights might be warranted. Reforms to realize these sorts of arrangements would need to be devised by people with intimate knowledge of what is required, what the existing arrangements are, and how they might be adjusted. They would need people working from both within and outside the sector, across knowledge and policy boundaries and across institutions and organizations.

In this complicated yet potentially productive context, attempts by the tourism and recreation sector to alter existing policy settings, or reclaim already assigned policy ground on water rights and allocations, are unlikely to be achieved by conventional means. Relying on first doing the research for input to an undetermined policy process risks commissioning research that addresses the wrong questions. Simply lobbying politicians can be equally futile. Nor can the desired outcomes be achieved by staking out the sector's turf with purportedly objective knowledge demonstrating, for example, the greater value of tourism in comparison with other sectors, such as agriculture. While it is essential to calculate the economic value and importance of the sector (NLTS Steering Committee 2009; STCRC 2008) to put its water issues on the policy agenda, it is not sufficient to bring about the requisite changes in water policy to accommodate the tourism and recreation sector's water needs. Clearly, a better way forward is needed.

This chapter argues that there *is* a better way, and it centers on *how* knowledge is produced, which can influence *what* knowledge is produced, and in turn can influence the willingness and capacity of policymakers or end users to put that knowledge into action (Cash et al. 2006). To this end, this chapter proposes a coproduction knowledge governance model. Coproduction can be broadly conceived as a process of making knowledge and policy together (Jasanoff 1990, 2004). It would involve the tourism and recreation sector engaging in reciprocal dialogue, collaboration, and negotiation with, for example, scientists, modelers, economists, other water users, government agencies, policymakers, or civil society actors for mutual outcomes.

With a view to demonstrating the coproduction model as a plausible alternative for harnessing knowledge, this chapter begins with an outline of the theoretical and collaborative governance context for the use of coproduction. It then explains two conceptual frameworks that have been proposed by Cash et al. (2006) as a means to facilitate coproduction, namely, boundary objects (Star and Griesemer 1989) and boundary organizations (Guston 2001). To begin demonstrating how this model is relevant to the tourism and recreation sector, the model is then applied to the case studies presented in the three previous chapters. Next, to examine how coproduction, boundary objects, and boundary organizations have been used in other contexts, two case analyses are presented: water management in agriculture on the Great Plains of the United States and the development of regional El Niño/Southern Oscillation forecasts for the Pacific and southern Africa. This section also discusses three knowledge attributes and four institutional functions maintained by Cash et al. (2006) as crucial for the success of coproduction and the linking of knowledge to action. This is followed by an examination of the case analyses, with conclusions drawn about the utility of this model for the tourism and recreation sector. The insights here cannot be interpreted as prescriptive but should be used as guiding principles or as a platform from which to launch further research (e.g., Kelly et al. 2006). For more detailed explorations of the components of this coproduction model, see Scott (2000), Cash and Buizer (2005), Cash et al. (2003), Folke et al. (2005), and de Loë et al. (2009).

COLLABORATING TO COPRODUCE KNOWLEDGE

Coproduction, for the purposes of discussion in this chapter, is underpinned by constructivist theories of knowledge that have emerged from the field of science and technology studies (STS) (Jasanoff 2004; Latour 1993). Often deployed in STS to critique the use of science in public policy (e.g., Duncan 2004; Jasanoff 1987, 1990; Shackley and Wynne 1995, 1996), coproduction has been used more recently as a framework to evaluate science–policy interactions (Jasanoff 2004). Useful for the tourism and recreation sector is work undertaken by researchers of knowledge systems for sustainable development who see coproduction, boundary objects, and boundary organizations as a means to encourage the collaboration of knowledge producers and knowledge users to build stronger links between knowledge and action (Cash et al. 2006; Kelly et al. 2006). Coproduction is in stark contrast to the conventional linear model of knowledge production for policy, which constitutes the domains of science and politics, or knowledge producers and knowledge users, as mutually exclusive (Jasanoff and Wynne 1998; Owens et al. 2006). Crucially, coproduction collapses the putative boundary between science and politics (Gieryn 1983; Jasanoff and Wynne 1998; Latour 1993). If appropriately deployed, coproduction could be a better way forward for the tourism and recreation sector to harness knowledge and renegotiate existing water policy settings.

Coproduction aligns with the shift in recent years from government to governance, whereby more collaborative approaches to policymaking have been called for and, in varying degrees, adopted (de Loë et al. 2009; Folke et al. 2005; Ostrom 1997; Pahl-Wostl et al. 2007; Scholz and Stiftel 2005). For example, the preceding case studies call for and illustrate the importance of opening up planning and decisionmaking to a broad range of perspectives. To see the spectrum of values embodied in recreation reflected in water policy, Chapter 9 calls for a cooperative governance approach, with the involvement of government agencies, nongovernmental organizations (NGOs), members of the community, and interested stakeholders. Likewise, the case study in Chapter 8 provides an instructive contrast between the confused governance arrangements that have given rise to the "look but do not touch" Perth Water and the governance arrangements embracing community engagement, collaboration, and negotiation that were the foundation of the vibrant Harlem River Park in New York.

It is increasingly recognized that the changing governance context also needs to extend to knowledge governance (Irwin and Wynne 1996; Jasanoff 2004; Latour 1993). This shift is evident in the sustainability and adaptive governance literature (e.g., Cash et al. 2003; de Loë et al. 2009; Folke et al. 2005; Owens et al. 2006; Pahl-Wostl et al. 2007). In this context, the tourism and recreation sector has an opportunity to do much more than calculate and proffer its market and nonmarket economic values in the hope of instigating water policy change. Rather, given its potential to augment full or partial complementarities with other sectors—that is, economic development, ecosystem services, and human well-being—it could be working to simultaneously harness knowledge and negotiate reforms for shared outcomes with other water users. Two concepts that have emerged from the field of STS that could facilitate engagement, communication, collaboration, and negotiation in coproduction are boundary objects and boundary organizations (Cash et al. 2006; Guston 2001; Star and Griesemer 1989).

Boundary Objects

According to Star and Griesemer (1989), the "boundary object" is a means by which people from "distinct social worlds" with different worldviews and values can cooperate. The boundary object concept originated in this much-cited 1989 article focused on the establishment of the Vertebrate Zoology Museum at the University of California–Berkeley between 1907 and 1939. The authors' central question was, how were cooperation, coherence, and credibility achieved in the midst of the divergent worldviews, practices, and vocabularies of professional and amateur museum collectors, with the need to arrive at conclusions that were acceptable to all? Their answer was the boundary object—a concept or something more material that brings together and accommodates divergent social worlds.

Star and Griesemer maintain that boundary objects can be "abstract or concrete" and describe them as "plastic enough to adapt to local needs and constraints of the several parties employing them, yet robust enough to maintain common identity across sites" (1989, 393). Boundary objects are all around us, in the form of maps, models, management plans, forecasts, policies, and treaties, to name a few. Star and Griesemer define a boundary object as "an object which lives in multiple social worlds and which has different identities in each" (1989, 409). Hence, its scope for multiple interpretations is an important characteristic. If appropriately chosen and deployed, boundary objects can facilitate cooperation, coherence, and credibility across knowledge and institutional boundaries. Notably, according to Star and Griesemer, the use of boundary objects is not intended to generate consensus. Rather, they mediate the alignment of mutual interests by allowing those involved to retain their varying perspectives while also contributing to building common ground. In short, boundary objects can be a means of mobilizing conflicting viewpoints for mutual outcomes. It will be shown in the case analyses to come that their success in application can be variable.

Boundary Organizations

Boundary organizations, like boundary objects, are used to bridge divergent social worlds, but they play a broader role by facilitating, if not institutionalizing, the use of boundary objects. Boundary organizations, as formal institutions or informal institutional collectives, can serve to translate ideas, vocabularies, practices, and worldviews across knowledge and institutional boundaries that ordinarily serve to isolate science and policy communities. Guston (1999) identifies three essentials for boundary organizations: they facilitate and validate the use of boundary objects, bring together science and policy actors as well as professional mediators to enable coproduction, and embody arrangements that ensure accountability on both sides of a boundary. According to Guston, boundary organizations need to be strategically positioned between divergent social worlds and, importantly, accountable to each of them.

Scott identifies boundary organizations as the "most important observation" of a review he conducted on the capacity of the European Environment Agency to disseminate its environmental research and sees boundary organizations as being able to "perform tasks that are useful to both sides, and involve people from both communities in their work, but play a distinctive role that would be difficult or impossible for organizations in either community to play" (2000, 5, 15).

Hence, boundary organizations are like conduits—they encourage communication, information, and ideas to flow. They can serve to mediate and translate across knowledge and policy boundaries and across institutions and organizations. Guston argues that rather than insulating itself from external political forces, the boundary organization's success is determined by its "being accountable and responsive to opposing, external authorities" (2001, 402). In other words, boundary organizations need to engender trust and goodwill internally and externally.

To begin to illustrate how these insights might be useful for the tourism and recreation sector, they can be applied to the case studies presented in Chapters 8 and 10. For example, the Boat Blue image discussed in Chapter 10 could be conceived as a boundary object. It was the centerpiece of the campaign led by the National Marine Manufacturers Association (NMMA) to overturn a court ruling that nullified an exemption for recreational vessels from regulation under the Clean Water Act in the United States. More than 60 different organizations identified with the Boat Blue theme, which helped mobilize considerable public, legislative, and institutional action. Furthermore, NMMA could be conceived as a boundary organization. It mobilized the Boat Blue boundary object. It successfully translated across knowledge and institutional boundaries the concerns and needs of recreational vessel owners to parties in the legislature, the US Environmental Protection Agency, civil society, and the broader public. It mediated divergent worldviews in the establishment of its coalition. However, it could be argued that NMMA was engaged predominantly in a one-way dialogue for a specific purpose rather than the reciprocal dialogue needed for coproduction.

The Boating Industry Association of South Australia (BIASA) provides a useful illustration of what reciprocal dialogue in coproduction might look like. As a boundary organization, BIASA operated vertically among state, national, and international jurisdictions and horizontally as an intermediary among its constituents, state agencies, and local governments. Via BIASA, concerns about lowering water levels for boating along the River Murray in Australia were conveyed to governments. Given its interest and navigational capacity, BIASA was invited by the state government to become involved in the production of knowledge—a survey of river accessibility for recreation and tourism. Instigating reciprocal dialogue, BIASA engaged with government agencies at the various levels to identify issues to be considered as part of its survey work and invited government representatives to be involved. After conducting the survey for a number of years, BIASA created its own boundary object, a publication titled *South Australia's Waters: An Atlas and Guide.* The durability of BIASA as a boundary organization and its boundary object is demonstrated by the continued involvement of BIASA with all tiers of government in Australia, NGOs, and international bodies. It also continues to be a source of on-ground knowledge and advice for working parties, task forces, and policy committees.

Finally, the initial master plan and subsequent management plan that facilitated the reinvention of the Harlem River Park in New York could also be conceived as boundary objects for their role in creating common ground and encouraging cooperation in the site's redevelopment. This case study in Chapter 8 underscores what is possible when responsibility, accountability, and resources rest with one body—a boundary organization, in this case the Harlem River Park Task Force—with a mandate to engage, communicate, negotiate, and collaborate broadly.

COPRODUCTION IN ACTION

This section looks at how boundary objects and boundary organizations have facilitated coproduction in practice in other contexts. The first case is water management for irrigated agriculture on the US Great Plains. Cash (2001) examines the role of the agricultural extension system as a boundary organization in helping farmers manage their irrigation practices with a depleting aquifer water source. He found that county extension agents and specialists using socio-economic, hydrogeologic, and cropping computer models involved farmers in model development and collection and use of data. The following quote from an interview conducted by Cash with a county extension agent illustrates the sort of facilitation, negotiation, and mediation work the agents were involved in as boundary operators:

> There was a question of a policy [regulatory] change from the Ground Water Management District, and the producers [farmers] were questioning whether the policy was going to affect them adversely or not. And so it was a producer-driven need for an answer, to give them some credible knowledge to make a decision on whether or not they wanted that [new] policy in place. And so, as the agents, we contacted the university to find who was doing the study.... We got the department heads out there... the head of economics ... and a couple of others. And we sat down with the members of the water board. ... We sent letters to producers and got a group of producers together, and all of us sat down and hashed out what we would like to see done here. And the university went back and set up the model, and started working on the model, and then we started putting the baseline data together. ... And it was a back and forth thing for several years getting it done because it was a rather involved model. (Cash 2001, 441; brackets in original)

This is a description of the coproduction of knowledge via a boundary object facilitated by a boundary organization. The knowledge users—that is, the farmers—were actively involved with knowledge producers, in this case the modelers, to create knowledge that became robust in the eyes of the farmers, modelers, and regulator. How was this achieved? With the farmers' participation, the modelers had insight into on-ground issues and conditions. They also had access to local data and on-ground knowledge for identifying necessary model parameters. This gave their model outputs credibility for all involved and allowed the modelers to produce policy-relevant decisionmaking tools. The farmers and water managers were able to run scenarios they trusted, as they had been involved with the modelers in the development of the model and the collection of data inputs. In their role as mediators of disparate social worlds, the county extension agents brought together a broad range of actors and translated their needs and goals across knowledge, policy, and institutional boundaries that otherwise would have isolated farmers, modelers, and the water board. In summary, Cash concludes that "neither community could have produced a model that was relevant and perceived as being scientifically sound without the other's participation. The county agent, in this case, acted as facilitator across the boundary between these two groups" (2001, 441). Hence, the coproduction of knowledge and policy was facilitated by the use of the predictive model, a boundary object, and the county extension system, a boundary organization.

The second case focuses on the translation of global El Niño/Southern Oscillation (ENSO) research into regional forecasts for the Pacific and southern Africa. The acronym ENSO represents the atmospheric pressure and oceanic temperature phenomenon across the Pacific Ocean that periodically, depending on ocean temperatures and atmospheric conditions, leads to low rainfall and drought conditions in countries on either side of the Pacific and farther afield. A number of institutions around the world have been involved in understanding and predicting the onset, frequency, and intensity of ENSO events. This global information is tailored for local needs by regional bodies in efforts to minimize the risks of loss of life, livelihoods, and infrastructure occasioned by unpredicted and unplanned-for weather events (Cash et al. 2006).

A comparative institutional study undertaken by Cash et al. (2006) examined two regional bodies: the Pacific ENSO Applications Center (PEAC) and the Southern African Development Community (SADC) Drought Monitoring Center (DMC). PEAC was established in 1994 and encompassed Hawaii and the United States Affiliated Pacific Islands (USAPI). Its mission was to "conduct research and forecasting for the benefit of the USAPI and the islands' various economic, environmental, and human services sectors" (Cash et al. 2006, 476). Illustrative of its positioning as a boundary organization, the institutions that PEAC's establishment brought together included the US National Oceanic and Atmospheric Administration (NOAA), its Office of Global Programs, and its National Weather Service Pacific Region; the University of Hawaii, its Social Science Research Institute, and its School of Ocean and Earth Science and Technology; the University of Guam and its Water and Energy Research Institute; and the Pacific Basin Development Council (Cash et al. 2006).

DMC in Zimbabwe encompassed Southern Africa. Its establishment was led by SADC, which had developed relationships with NOAA; the World Meteorological Organization; World Bank; National Meteorological Services of Southern Africa member countries; United Nations Development Programme; International Research Institute for Climate Prediction; United Kingdom Meteorological Office; and SADC's Regional Early Warning Unit, Regional

Remote Sensing Unit, and Famine Early Warning Systems Network. The purpose of DMC was to take advantage of the emerging ENSO science, which had the potential to improve planning for Southern Africa to avoid the disastrous consequences of drought and famine (Cash et al. 2006).

Clearly, each regional body forged relationships with many organizations and institutions across a range of knowledge and institutional boundaries to develop their forecasts. Retrospectively, Cash and colleagues conceived of each body as a boundary organization on the basis that each had served as an intermediary between the respective knowledge producer and user communities. In their 2006 study, Cash *et al.,* examined and compared the mechanisms used by these bodies to bring together actors and groups from different fields, how information was translated across boundaries, the effectiveness of these mechanisms in collaboration, and the methods these entities used for resolving conflicts.

Cash et al. (2006) maintain that the knowledge attributes of salience, credibility, and legitimacy (Cash and Buizer 2005; Clark and Dickson 1999; Jasanoff 1990) are essential for successfully linking knowledge to policy action. Salience is about relevance. Does the knowledge outcome answer the right question, and is it in a useful form and provided at a useful time? Salience is indispensable to end users. Credibility is about technical adequacy. Have appropriate methods been used? How were data obtained? What analysis has been applied? Are the conclusions robust? Issues such as these are particularly important for the scientific community, but they are for end users as well. Legitimacy is about fairness. Is the knowledge production process fair and open? Are there mechanisms to appropriately facilitate the expression of values and the resolution of conflicts? These issues are important for end users and the wider community (Cash et al. 2006; Clark and Dickson 1999; Scott 2000).

It is evident that attributes of salience, credibility, and legitimacy were accomplished by the Great Plains agricultural extension system. At the instigation of farmers and via the extension agent, the farmers' involvement with modelers ensured that the model, the boundary object, generated information they needed and wanted. This gave the model and its outputs salience. The farmers' contribution to the process also gave the modelers local knowledge and data. The modelers gave the farmers insight into how the model worked, what it could do, and how it could help them. The modelers were able to incorporate the farmers' inputs without compromising technical adequacy and their credibility. The county extension agents acted to mediate between the parties so that coproduction proceeded with cooperation and engendered legitimacy.

Importantly, the attributes of salience, credibility, and legitimacy are interdependent—a shift in one can shift another. Cash et al., argue that "threshold levels of salience, credibility, and legitimacy" need to be maintained "while managing trade-offs between them" (2006, 468). For example, for farmers, the salience of a predictive model might be heightened if its variables were altered in some way, but if this compromised the technical adequacy of the model, its credibility would be diminished for the modelers. If no mediation process existed to allow farmers and modelers to air their concerns and negotiate a resolution, the legitimacy of the knowledge outputs would be diminished. It is the role of a boundary organization to facilitate the resolution of these sorts of issues. Conversely, the production of knowledge in isolation from end users might enhance credibility among scientists, but salience would be diminished if it answered the wrong questions because of a lack of engagement and collaboration. Additionally, legitimacy would founder if end users perceived the knowledge-producing process as unfair. A well-functioning boundary organization would ensure that engagement and collaboration occurred. In summary, all three criteria are interlinked, and attempts to boost one can come at the expense of another (Cash et al. 2006). Hence, the challenge in processes of coproduction is to ensure that all three are maintained at adequate levels.

According to Cash et al. (2006), these attributes can be managed and held at adequate levels by boundary organizations convening, translating, collaborating, and mediating, across knowledge and policy boundaries, and across institutions and organizations. Cash and colleagues assessed the effectiveness of PEAC and DMC to produce salient, credible, and legitimate knowledge using these four institutional functions, described as follows:

- *Convening* involves bringing producers of knowledge and end users together face-to-face to facilitate dialogue and build trust.
- *Translating* requires communication, assisted by boundary organization intermediaries, to translate language, jargon, assumptions, methods, world-views, and practices to facilitate the breaking down of barriers and flow of information and ideas.
- *Collaborating* means putting actors from different social worlds to work on a boundary object, such as a model or forecast.
- *Mediating* is about resolving conflicts that arise when issues or divergent ideas, values, or interests collide. It involves open evaluation and mediation by people who are respected and trusted by those involved.

Following is an outline of the research reported by Cash et al. (2006) to assess how the four institutional functions were applied by the boundary organizations of PEAC and DMC, and the divergent outcomes for salience, credibility, and legitimacy arising from their differential application.

Boundary Organization Convening

To develop its regional forecast, PEAC organized events to bring together three groups—scientists (e.g., climate scientists, hydrologists, epidemiologists, and economists), forecasters (e.g., meteorologists), and end users (e.g., water and fisheries managers, emergency service managers, and government officials)—who started working together in the initial planning stages of PEAC. According to Cash *et al.,*, representatives of these groups were involved as "joint collaborators in designing the scope of an ENSO research and applications system in the Pacific" (2006, 477). This move placed these actors within the boundary organization and spread decisionmaking and accountability responsibilities across the three groups. Overall, PEAC had a responsibility to engage broadly, as its funding was contingent on meeting its founding socioeconomic and environmental mission. The involvement within the organization of the representatives of the three groups also meant that PEAC was obliged to facilitate dialogue and work on building trust among them.

The DMC forecast started with meteorologists coming together in workshops to establish parameters for individual countries, which set the groundwork for the DMC's Southern African Regional Climate Outlook Forum (SARCOF). This forum was held twice a year and, like PEAC, brought together scientists, forecasters, and decisionmakers from across its region. Convening at SARCOF involved the forecasters hearing presentations on issues and information needs from end users (e.g., the World Food Program, hydropower planners, and researchers). The scientists then arranged to "meet among themselves to iron out differences between their national forecasts," and after their deliberations, they "present[ed] their consensus forecast to the others at the meeting" (Cash et al. 2006, 479). This application of convening did not demonstrate to end users that those involved had listened to or considered their issues. Nor was reciprocal dialogue facilitated to build trust among the parties. This convening generated credibility for the forecast within the scientific community, but it came at the expense of salience and legitimacy for end users.

Boundary Organization Translation

As well as translating the forecast information into the languages spoken in different countries, PEAC used scientists who were adept at translating scientific jargon into lay terms and cognizant of the needs of a range of end users. For example, probabilities were translated into a more user-friendly language. In addition, the timing of the release of the forecast was negotiated among all the parties to ensure that it was not so early that it would be forgotten or so late that it would leave no time to prepare. This gave PEAC's forecast salience and legitimacy while maintaining sufficient levels of credibility. Given its predominance of meteorologists, DMC paid little attention to translation at SARCOF. Probabilities were deemed acceptable for communicating to end users. Thus, the credibility of this forecast came at the expense of salience and legitimacy for end users.

Boundary Organization Collaboration

PEAC encouraged and facilitated multiple actors to work on the coproduction of the forecast, the boundary object. Each contributor had a stake in the outputs and gained insight into the perspectives of others:

> Each actor clearly benefited in a different way from the collaboration: receiving information on predicted rainfall for an agricultural-extension officer; hearing warnings of impending storms for an emergency manager; learning predictions of where fish might be for the fishing industry; and producing a publication in a peer-reviewed journal for a scientist. Though the forecasts had different value and meaning for each of these actors (an

important part of being a boundary object), PEAC was able to coordinate and mediate activities such that there was enough overlapping meaning that a robust forecast could be produced. (Cash et al. 2006, 480)

The DMC did not directly involve end users in developing its forecast. The following feedback Cash and colleagues received through personal communication from a 2002 SARCOF participant illustrates the lack of engagement, its implications for end users, and the knowledge attributes:

> First, there was a sharp divide between forecasters and forecast users. There were no users invited to the consensus forecast group, and no climate expert was involved in any of the four users' working groups (health, food and agriculture, water and energy, and disaster management). Even though the users' group presentations had a lot of demands in common, no climate person made any effort to share perspectives about the feasibility of satisfying users' needs. For example, beginning and duration of rainy season was clearly something desired, but users left with no knowledge of whether climate scientists can (or want to) provide that information to them. (Cash et al. 2006, 481–482)

Yet again, although the forecast might have had credibility among the scientists and forecasters, it lacked salience and legitimacy for those that needed and wanted to use it.

Boundary Organization Mediation

Because the PEAC parties were actively involved in coproduction, conflicting values and interests were an issue of concern from the outset, so mediation arrangements were put in place. The following comments from a University of Hawaii social science researcher in this role illustrate how productive such processes can be in dispelling myths about knowledge needs:

> [We] facilitated a dialogue between the scientists and decision-makers about … the beliefs of scientists about information and about certainty and uncertainty and probability and then the beliefs and needs of bureaucrats about the same thing. And we had the scientists characterize the bureaucrats and the bureaucrats the scientists, and then we brought them all back in the room together and said, this is what they said about you, and this is what they said … and it is interesting that … everybody thought the other one needed certainty. (Cash et al. 2006, 482)

The provision of active mediation by trusted intermediaries created common ground among the parties and established legitimacy and salience for the PEAC forecast. For the DMC, on the other hand, mediation was complex and influenced by broader political issues in Zimbabwe at the time. The DMC did not have access to the needed human resources or embody the institutional mechanisms for mediating conflicts. End users were not directly involved in the forecasting process. Dialogue and negotiations occurred within the scientific community

Table 11.1. *Summary of the achievement of knowledge attributes derived from the application of institutional functions by PEAC and DMC*

	Salience	Credibility	Legitimacy
PEAC			
Convening	√	√	√
Translation	√	√	√
Collaboration	√	√	√
Mediation	√	√	√
DMC			
Convening		√	
Translation		√	
Collaboration		√	
Mediation		√	

behind closed doors at SARCOF. Hence, although DMC's mediation occurred among scientists, which contributed to the credibility of the forecast in the scientific community, it did not contribute to the legitimacy and salience of the forecast for end users.

Table 11.1 summarizes the success of each boundary organization, PEAC and DMC, in carrying out the institutional functions necessary to produce salient, credible, and legitimate knowledge for end users.

It can be seen that PEAC achieved salience, credibility, and legitimacy in respect to all four institutional functions. Accordingly, PEAC would be classed as a more successful boundary organization than DMC, and PEAC's forecast would be deemed to be a more useful and durable boundary object, as it met the needs of its stakeholders across the three knowledge attributes. Clearly, the credibility of the DMC forecast came at the expense of its legitimacy and salience with end users and, consequently, its likelihood of application on the ground (Cash et al. 2006).

CONCLUSIONS

The Great Plains water management case showed that boundary objects and boundary organizations can accommodate divergent worldviews and facilitate coproduction. The county extension system, as a boundary organization, was positioned between knowledge producers and knowledge users, and it was accountable to both in terms of funding and mandated outcomes. It facilitated the flow of communication and information by creating a bridge among farmers, government agencies, scientists, and modelers. Metaphorically, it built a two-way bridge to allow the flow of communication, information, and negotiation across knowledge, institutional, and organization boundaries. It was not a one-way bridge, as could be argued in the case of the NMMA with its Boat Blue campaign. The predictive model that simulated water depletion scenarios on the Great Plains—the boundary object—introduced a common focus, and with all parties involved in its development and use, it had legitimacy, collectively and individually.

Notably, the roles of farmers and modelers became interchangeable in terms of who was producing knowledge and who was using it. For example, because of the farmers' local lay knowledge and essential data inputs, they became knowledge producers, and the modelers took on the role of knowledge users (Duncan 2008; MacKenzie 1990; Shackley and Wynne 1995). This interchangeability of roles enhanced the salience and legitimacy of the knowledge outcomes without compromising their credibility.

The ENSO case illustrates that the utility and durability of boundary objects and boundary organizations can be variable. PEAC was a more effective boundary organization than DMC, and it mobilized a boundary object that was more useful to end users than was DMC's. Hence, PEAC was able to forge a strong link between knowledge and policy action, much more so than DMC. Both case analyses underscore the importance of managing the salience and legitimacy of knowledge if there is an expectation for it to be applied on the ground by end users. Both cases also show that the institutional functions of convening, translating, collaborating, and mediating are crucial for balancing salience, credibility, and legitimacy and, consequently, linking knowledge with policy action.

The case analyses underscore the extent to which the scientific credibility of knowledge can come at the price of its relevance and legitimacy in socioecological decisionmaking. This is an enduring tension. Ordinarily, it isolates knowledge producers and knowledge users with putative imperatives of scientific authority and policy purity (Jasanoff and Wynne 1998). It presents considerable challenges to knowledge-making and water policy action. Importantly, the case analyses have also shown that reciprocal dialogue, collaboration, and negotiation between knowledge producers and knowledge users can be achieved without compromising the credibility of knowledge outcomes or policy actions. In other words, scientists do not have to relinquish their scientific credibility to accommodate end users in the production of policy-relevant knowledge. Hence, a recognition of the need to balance the knowledge attributes of salience, credibility, and legitimacy in the application of the institutional functions of convening, translating, collaborating, and mediating goes some way toward addressing this tension. This coproduction model does so not by dispensing with the need for establishing credibility but by enhancing it with the dimensions of salience and legitimacy. Indeed, it would appear that with coproduction, credibility can be substantially bolstered.

With institutional functions in place, the changing governance context of water policy to more collaborative approaches can and should extend to knowledge governance. Water management on the Great Plains of the United States and the development of ENSO forecasts for the Pacific and Southern Africa show the utility of coproduction, boundary objects, and boundary organizations. Together with this chapter's interpretations of the case studies presented in the earlier chapters in light of the coproduction model, it is apparent that useful outcomes are possible for the tourism and recreation sector.

Shifts in knowledge governance have import for the tourism industry more broadly. The challenges for this industry across a range of areas to transfer and utilize knowledge have been acknowledged as considerable by Cooper (2006) and Shaw and Williams (2009). Forecasting to predict the spending, numbers, and origins of tourists is vital for the industry to manage economic risk (Fleetwood 2004), and it is reliant on knowledge derived from research and predictive modeling (NLTS Steering Committee 2009; Song and Li 2008). Considerable public resources are dedicated to gathering, collating, and modeling data and disseminating forecasts and analyses (Fleetwood 2004). Notably, a succession of strategies for Australia's tourism industry have sought to address an apparent disconnect among forecasts, their dissemination, and the payoff by way of tourism dollars (Commonwealth of Australia 2003; Fleetwood 2004; NLTS Steering Committee 2009). The coproduction model, with its emphasis on creating knowledge *with* rather than *for* end users, could be a better way forward for the tourism industry as well.

Problematically, it is possible that before the coproduction model is seriously entertained, it could be scuttled on the basis of the valid concern that the knowledge derived therefrom could become "crudely instrumental" (Owens et al. 2006, 636). The issue of who governs scientific knowledge and how it should be deployed is often cleaved between the poles of the "scientization of politics" and the "politicization of science" (Guston 2001, 405). The former would see scientific conclusions driving policymaking that might not take account of cultural, social, and economic factors on the ground. The latter would see the scope of scientific endeavor narrowed down and even more ad hoc than is currently the case (Dovers and Wild River 2003, 3). However, Guston sees coproduction and boundary organizations as the common ground needed to navigate an alternative path. "The politicization of science is undoubtedly a slippery slope. But so is the scientization of politics. The boundary organization does not slide down either slope because it is tethered to both, suspended by the coproduction of mutual interests" (2001, 405). Cash et al., argue that coproduction is a means to link knowledge and decisionmaking "in ways that are more socially embedded and that attempt to better balance economic, cultural, and social needs" (2006, 466).

Duncan and Hay (2007) raise the concern that the balance discourse of sustainable development to which Cash and colleagues refer has the potential not only to undermine the protection of the environment, but also to fast-track its trade-off for short-term social and economic gains. Despite the arguments of Guston (1999, 2001) and Cash et al. (2006), scientists, policymakers, and a range of civil society actors who would need to be involved in recalibrating water policy, who might see themselves as advocates for the environment or unwilling to concede any more environmental ground, would want to be sure that their involvement in coproduction had no chance of contributing to such problematic outcomes. Notwithstanding this important issue, the proposed coproduction knowledge governance model provides the tourism and recreation sector with a road map to reconceptualize, renegotiate, and redraw its existing knowledge and policy boundaries in terms of water policy. It also provides a means to harness the complementarities this sector has with other water uses and users and to potentially resolve conflicts arising from partial complementarities.

Given its limited knowledge base, its diffuse nature, and the complexity of its water needs, the tourism and recreation sector cannot rely on conventional modes of knowledge and policymaking to seek and secure adjustments to existing water policy settings. The sector needs to do more than calculate and promote its market and nonmarket economic value. Although the latter is necessary for setting a policy agenda, it is not sufficient to bring about the requisite changes in water policy to accommodate the tourism and recreation sector's water needs. Hence, an alternative knowledge governance model of coproduction has been proposed. It centers on *how* knowledge is produced, which can influence *what* knowledge is produced, which in turn can influence the willingness and capability of policymakers or end users to put that knowledge into action.

Coproduction is a useful knowledge governance model, not only for its potential to accommodate the complex needs of the tourism and recreation sector, but also for its possibilities to find ways to elicit cooperation from parties that could be expected to initially object to proposals to recalibrate water policy settings. In its favor are the inherent complementarities (full and partial) that the tourism and recreation sector holds, which could be leveraged with other water users to consolidate economic development, ecosystem services, and human health and well-being. These complementarities represent opportunities for the sector to pursue coincident interests in the coproduction of knowledge for mutual water outcomes. Putting coproduction into practice can be aided substantially by the use of boundary objects as well as boundary organizations that implement the institutional functions of convening, translating, collaborating, and mediating to balance salience, credibility, and legitimacy. Given the mediated and negotiated nature of coproduced knowledge, it is unlikely that the tourism and recreation sector would achieve all that it seeks to attain, but it is possible that this model of knowledge governance could take it a considerable way toward resetting existing water policy arrangements.

REFERENCES

ABS (Australian Bureau of Statistics). 2009. *Australian National Accounts: Tourism Satellite Account, 2008–09.* cat.5249.0. Canberra: Australian Government Publishing Service.

Cash, D.W. 2001. In order to aid in diffusing useful and practical information: Agricultural extension and boundary organizations. *Science, Technology & Human Values* 26 (4): 431–453.

Cash, D.W., J.C. Borck, and A.G. Patt. 2006. Countering the loading-dock approach to linking science and decision making: Comparative analysis of El Niño/Southern Oscillation (ENSO) forecasting systems. *Science, Technology & Human Values* 31 (4): 465–494.

Cash, D.W., and J. Buizer. 2005. Knowledge-action systems for seasonal to interannual climate forecasting: Summary of a workshop. *Roundtable on Science and Technology for Sustainability.* Canberra: National Research Council.

Cash, D.W., W.C. Clark, F. Alcock, N.M. Dickson, N. Eckley, D.H. Guston, J. Jager, and R.B. Mitchell. 2003. Knowledge systems for sustainable development. *Proceedings of the National Academy of Science of the United States of America* 100 (1): 8086–8091.

Clark, W., and N. Dickson. 1999. The global environmental project: Learning from efforts to link science and policy in an interdependent world. *Acclimations* 8: 6–7.

Commonwealth of Australia. 2003. *Australian Government Tourism White Paper: A Medium to Long Term Strategy for Tourism.* Canberra: Commonwealth of Australia.

Cooper, C. 2006. Knowledge management and tourism. *Annals of Tourism Research* 33 (1): 47–64.

de Loë, R.D., D. Armitage, S. Davidson, and L. Moraru. 2009. *From Government to Governance: A State-of-the-Art Review of Environmental Governance.* final report, prepared for Alberta Environment, Environmental Stewardship, Environmental Relations. Guelph, Ontario: Rob de Loë Consulting Services.

Dovers, S., and S. Wild River. 2003. *Managing Australia's Environment.* Sydney: Federation Press.

Duncan, R. 2004. Science Narratives: The Construction, Mobilisation and Validation of Hydro Tasmania's Case for Basslink, PhD thesis, Hobart, School of Geography and Environmental Studies, University of Tasmania.

———. 2008. Problematic practice in integrated impact assessment: The role of consultants and predictive computer models in burying uncertainty. *Impact Assessment and Project Appraisal* 26 (1): 53–66.

Duncan, R., and P. Hay. 2007. A question of balance in integrated impact assessment: Negotiating away the environmental interest in Australia's Basslink project. *Journal of Environmental Assessment Policy and Management* 9 (3): 273–297.

Fleetwood, S. 2004. Current Developments in Expansion of Australia's Tourism Data. paper presented at 7th International Forum on Tourism Statistics, June 9-11, 2004, Stockholm.

Folke, C., T. Hahn, P. Olsson, and J. Norberg. 2005. Adaptive governance of social-ecological systems. *Annual Review of Environment and Resources* 30: 441–473.

Gieryn, T. 1983. Boundary-work and the demarcation of science from non-science: Strains and interests in professional ideologies of scientists. *American Sociological Review* 48: 781–795.

Guston, D.H. 1999. Stabilizing the boundary between politics and science: The role of the Office of Technology Transfer as a boundary organization. *Social Studies of Science* 29 (1): 87–112.

———, 2001. Boundary organizations in environmental policy and science: An introduction. *Science, Technology & Human Values* 26 (4): 399–408.

Irwin, A., and B. Wynne. 1996. Introduction. In *Misunderstanding Science? The Public Reconstruction of Science and Technology,* edited by A. Irwin and B. Wynne. Cambridge, UK: Cambridge University Press, pp. 1–17.

Jasanoff, S. 1987. Contested boundaries in policy-relevant science. *Social Studies of Science* 17: 195–230.

———, 1990. *The Fifth Branch: Science Advisers as Policymakers.* Cambridge, MA: Harvard University Press.

———, 2004. Ordering knowledge, ordering society. In *States of Knowledge: The co-production of science and social order,* edited by S. Jasanoff. London: Routledge, pp. 13–45.

Jasanoff, S., and B. Wynne. 1998. Science in decisionmaking. In *Human Choice and Climate Change,* Volume 1: *The Societal Framework,* edited by S. Rayner and E.L. Malone. Columbus, OH: Battelle Press, pp. 1–87.

Kelly, T., J. Reid, and I. Valentine. 2006. Enhancing the utility of science: Exploring the linkages between a science provider and their end-users in New Zealand. *Australian Journal of Experimental Agriculture* 46: 1425–1432.

Latour, B. 1993. *We Have Never Been Modern.* Translated by Catherine Porter. Cambridge, MA: Harvard University Press.

MacKenzie, D. 1990. *Inventing Accuracy: A Historical Sociology of Nuclear Missile Guidance.* Cambridge, MA: MIT Press.

NLTS (National Long-Term Tourism Strategy) Steering Committee. 2008. *The Jackson Report: The National Long-Term Tourism Strategy.* Canberra: Commonwealth of Australia.

Ostrom, E. 1997. Crossing the Great Divide: Coproduction, synergy, and development. In *Global, Area, and International Archive,* eScholarship, University of California.

Owens, S., J. Petts, and H. Bulkeley. 2006. Boundary work: knowledge policy, and the urban environment. *Environment and Planning C: Government and Policy* 24: 633–643.

Pahl-Wostl, C., M. Craps, A. Dewulf, E. Mostert, D. Tabara, and T. Taillieu. 2007. Social learning and water resources management. *Ecology and Society* 12 (2): 5.

Pigram, J.J. 2006. *Australia's Water Resources: From Use to Management.* Collingwood: CSIRO Publishing: 173–191.

Scholz, J.T., and B. Stiftel. 2005. *Adaptive Governance and Water Conflict: New Institutions for Collaborative Planning.* Washington, DC: Resources for the Future.

Scott, A. 2000. *The Dissemination of the Results of Environmental Research: A Scoping Report for the European Environment Agency.* Copenhagen: European Environment Agency.

Shackley, S., and B. Wynne. 1995. Integrating knowledges for climate change: Pyramids, nets and uncertainties. *Global Environmental Change* B (2): 113–126.

———, 1996. Representing uncertainty in global climate change science and policy: Boundary-ordering devices and authority. *Science, Technology & Human Values* 21 (3): 275–302.

Shaw, G., and A. Williams. 2009. Knowledge transfer and management in tourism organisations: An emerging research agenda. *Tourism Management* 30: 325–335.

Song, H., and G. Li. 2008. Tourism demand modelling and forecasting: A review of recent research. *Tourism Management* 29: 203–220.

Star, S.L., and J. Griesemer. 1989. Institutional ecology, 'translations' and boundary objects: Amateurs and professionals in Berkeley's Museum of Vertebrate Zoology, 1907-39. *Social Studies of Science* 19 (3): 387–420.

STCRC (Sustainable Tourism Co-operative Research Centre). 2008. *Tourism Satellite Accounts 2006–07: Summary Spreadsheets.* Gold Coast, Queensland: CRC for Sustainable Tourism Pty. Ltd.

PART IV

TOURISTS AND URBAN WATER
AND LESSONS FOR THE FUTURE

CHAPTER 12

The Use of Potable Water by Tourists: Accounting for Behavioral Differences

Bethany Cooper

T he previous chapters noted that numerous countries are approaching water crises that have largely been a function of population growth, industrialization, and expansion of irrigated agriculture. Recent research has also emphasized the strain that conventional tourism has on water resources around the world (see, e.g., Andereck 1995; Green et al. 1990; Honey 2001). The United Nations Commission on Sustainable Development (2010) notes that in some locations, an influx of tourists can more than double water demands, placing considerable stress on water infrastructure and water resources generally. Despite this impact, little empirical research has been conducted on the determinants of tourist water demand and related behavior. The result is that governments and local communities have limited information by which to anticipate tourists' demand for water, and they know even less about how to shape and guide tourist behavior (UNCSD 2010).

What is known is that individuals who may generally be model citizens are inclined to behave differently when vacationing (see, e.g., Bergin-Seers and Mair 2008; Wearing et al. 2002). These behavioral differences may encapsulate an altered concern for risk, and increasing evidence shows that water-using behavior varies when people are recreating or on vacation. Tourists also may respond to different motivators or triggers in framing their behavior (Gnoth 1997). Therefore, the appropriate measures to influence tourist behavior may be substantially different from those that are effective for locals.

In the context of water shortages, the behavior of tourists has significant implications for water infrastructure provision, pricing of water, and appropriate accreditation and incentives, just to name a few. This chapter explores the unique behavioral dimensions of tourists and their effects on key components of water policy. It begins with a brief review of the impacts of tourism on water resources from both international and national perspectives. It then examines tourist behavior from a theoretical perspective and looks at the tendency of individuals to behave differently when taking on the role of tourists. Following this is an investigation of the theoretical underpinnings of compliance behavior, drawing on the literature from the disciplines of psychology, sociology, and economics. Finally, a framework is presented that could be used to assist in developing more effective policy surrounding tourists' compliance behavior.

THE IMPACTS OF TOURISM ON WATER RESOURCES

Although tourism is one of the largest industries in the world and is often perceived as a "clean sector," it has a serious environmental footprint (UNESCO 2006). This section looks at the impacts of tourism on water resources at the international and national levels.

International Level

Gossling (2000) suggests that a number of developing countries have concentrated on tourism since the 1960s to create additional jobs and income, raise foreign exchange earnings, and diversify their economies. Although the economic benefits are evident, increased tourism presents increased pressure on natural resources and has potentially detrimental impacts. For instance, island states and coastal zones in the tropics have experienced substantial development in tourist infrastructure (see, e.g., Miller and Auyong 1991; Rodriguez 1981), which presents significant environmental problems (Gossling 2000). Small limestone island nations, such as Bermuda, the Cayman Islands, and the Bahama Islands, rely solely on rainfall for freshwater supply (Jones and Banner 1998; UN 1995), and these islands are often already overexploited. In drier regions like the Mediterranean, the issue of water scarcity is of particular concern because of the hot climate and tourists' tendency to use more water (UNESCO 2010). De Stefano (2004), who has investigated the impact of tourism on freshwater resources in the Mediterranean, suggests that tourism is a major contributing factor to the degradation of water ecosystems. Consequently, it is affecting not only habitats, but also human communities.

Hotels and their guests consume vast quantities of water. In Israel, water use by hotels along the Jordan is considered to be contributing to the drying of the Dead Sea (UNESCO 2010). Moreover, tourism has often been identified as a driving force of groundwater use. For example, in the Balearic Islands, in the western Mediterranean Sea and some coastal regions of Spain, saltwater intrusion has occurred largely as a result of the overuse of groundwater by the tourist industry (German Federal Agency for Nature Conservation 1997).

Despite the often limited availability of fresh groundwater, it remains one of the most sought-after resources by the tourist industry in coastal zones, highlighting the implicit risk of overexploitation. In such situations, reducing water consumption and preventing encroachment into wetlands can be achieved only if the tourism industry and individual tourists are prepared to adopt water-saving policies and behavior. Thus, although some progress has been made in terms of resource consumption, waste management, and transport optimization (Gotz et al. 2002; Upham 2001; VISIT 2005; WTTC 2007), much remains to be learned from a demand-side perspective to facilitate the design of effective policy.

National Level

Australia as a tourism destination is typically promoted for its coastal landscapes; recreational activities, such as water-based sports; and the amenity of many of its urban centers. However, the extended drought in recent years has meant that most urban centers have seen mandated restrictions constraining water uses and their timing. Against this background, it has been suggested that the tourism industry commonly overconsumes water resources for hotels, golf courses, swimming pools, and personal use by tourists (Honey 2001).

The challenge for policymakers is to ensure that tourism plays a positive long-term role in economic development in Australia, but in a way that is both environmentally and socially acceptable. Given the significant heterogeneity in the tourism industry, it seems inappropriate to devise blanket principles around sustainable tourism (Stabler and Goodall 1997). Bergin-Seers and Mair (2008) review the concept of sustainable tourism and suggest that it is prudent to investigate measures that may lead to the adoption of the principles of sustainable tourism among tourism firms. However, in addition to supply-side measures, there would appear to be value in exploring measures that may motivate more tourists to adopt eco-friendly behavior and, in particular, accept the need for responsible water-using behavior. An initial understanding of tourist behavior and trends in this field provides a useful starting point.

TOURIST BEHAVIOR

The study of tourist behavior stems from the field of consumer behavior, which has focused on modeling behavior since the 1960s (Engel et al. 1968; Howard and Sheth 1969). Consumer behavior concepts have provided significant insights into tourist behavior, taking account of findings in behavioral and cognitive psychology. Individuals are commonly driven by different factors when they are tourists, and their behavior has at times been categorized as irrational (Gnoth 1997).

Theoretical Underpinnings of Tourist Behavior

While tourism has been identified as a social and economic phenomenon, it is often suggested that tourist behavior should be considered primarily in psychological terms (Lewin 1942). Concepts from anthropology (Adler 1989), sociology, and sociopsychology (E. Cohen 1972, 1978; M. Cohen 1988; MacCannell 1992; Parrinello 1993) also contribute to understanding tourism in existential terms. Both sociology and psychology draw from the attitude construct for investigating and predicting tourist behavior. Attitudes are therefore a prominent concept in the development of a model for tourism motivation and development. Behavioral studies contribute to this literature by explaining the interplay of motivations that create the diversity of tourism demand (Ajzen 2001; Dann 1983; Wickens 2002).

Budeanu (2007) suggests that tourists are much less interested in adopting sustainable lifestyles or responsible tourism products than are corporate or government entities. Incentives and tools such as ecolabels, certification schemes, awards, and awareness and educational campaigns have been developed to encourage responsible tourist behavior, but their effectiveness is questionable (see, e.g., Chafe 2005; Martens and Spaargaren 2005). There is merit in investigating why there has been gradual adoption of environmentally benign tourism services and products but only minimal compliance with regulations to support sustainable behavior (Budeanu 2007). In the following sections, related strands of the tourist behavior literature are examined to gain insights into these issues.

Tourists and Risk-Taking Behavior

Risk is a common phenomenon, and every individual experiences it to some extent. Although many individuals try to avoid risk where possible, others appear to desire it. Pizam *et al.* define risk as "the possibility of experiencing a negative outcome" (2004, 251). In the past few decades, several researchers (Cook and McCleary 1983; Cossens and Gin 1994; Ewert 1994; Knopf 1983; Mansfeld 1992; Plog 1973; Roehl and Fesenmaier 1992; Smith 1990; Um and Crompton 1990) investigated the relationship between risk-taking and tourist behavior. In Plog's (1973) seminal research, leisure tourists are classified into two main personality types: allocentrics and psychocentrics. Allocentrics are tourists who seek out novelty vacation destinations that provide an escape from the confusion and boredom of life. These tourists are adventurous and relatively anxiety-free. Alternatively, psychocentrics usually desire familiar tourist destinations and are nonadventurous and take fewer risks. Plog's study recognizes allocentrics and psychocentrics as two extremes on a bell curve of tourist personalities.

Researchers have questioned whether individuals differ in the kinds of risk behavior they adopt (see, e.g., Pizam et al. 2004). It could also be questioned whether a general tendency exists for individuals to engage in risk behavior when they are on vacation. These risks may be physical, social, or psychological (Weber 2001). Moreover, different tourists may have various perceptions of what risks are integral to their vacation. For instance, Bentley et al. (2001) claim that tourists seek the risk associated with adventure tourism as part of the experience.

The literature on tourism risk-taking relates to tourist water use in two ways. First, urban water restrictions are usually in place to limit the risks of an urban population running dry. Arguably, the transient nature of tourist visitation and the proclivity for higher risk-taking reduce the perceived necessity of behavioral restrictions. Second, some risks attend noncompliance with water restriction. Although violation may result in penalties, tourists' preference for higher-risk behaviors could compound the difficulty of securing compliance.

Studies of tourists at both international and domestic destinations indicate increased risk-taking behavior, including greater frequency of alcohol and drug use and sexual activity (see, e.g., Downing et al. 2010). Other research in this domain supports the view that tourists are far less inclined to feel bound by conventions followed at home (see, e.g., Bellis et al. 2004).

The variation in the risk-taking behavior of individuals in the vacation setting leads to questions regarding other aspects of their behavior. For instance, are individuals as likely to be motivated to adopt sustainable consumption actions when on vacation as compared to when at home? Will they be inclined to be less conscious about their water consumption and thus take longer showers while on vacation?

Tourists and Sustainable Behavior

In addition to the literature on risk and tourist behavior, some work has been undertaken that directly relates attitudes and sustainability behavior to tourism. For example, in Firth and Hing's (1999) investigation of the attitudes and behaviors of guests in backpacker hostels in Byron Bay, 12% of all respondents claimed that they were reasonably environmentally conscious at home but abandoned this level of responsibility while on vacation. In a similar vein, Wearing et al. (2002) note that "a scenario begins to develop of individuals claiming to be concerned enough about the environment to factor these concerns into their choice of tourism products. However, under specific conditions, these same individuals, as tourists in a holiday [vacation] destination, seem unwilling to let concern for the natural environment affect their specific tourism purchasing behavior."

This supports the view that tourists may be prone to downplay environmental responsibility while on vacation. Arguably, tourists may need to be motivated by measures other than appeals to their environmental conscience, if the goal is to preserve or limit water use.

Notwithstanding the perspective that tourists may adopt eco-friendly behavior or behave consistently with regulations surrounding the consumption of natural resources, such as water restrictions, substantial evidence supports the idea that individuals tend to behave differently while on vacation. Thus, an individual's behavior at home might offer few clues about his or her behavior on vacation. In simple terms, setting modifies behavior.

Levitt and List (2007) propose a theoretical model that identifies three crucial differences between lab and field settings that can cause prosocial behavior to be quite different in the lab and the field. These differences are termed stakes, social norms, and scrutiny. This framework can be applied to a tourist behavior context and used to identify the crucial differences between vacation and home settings that influence sustainable behavior, and thus it can offer insights into alternative policy responses.

- *Stakes.* Levitt and List (2007) suggest that subjects in the laboratory are more likely to "play" with their money or resources. This highlights that the entitlement of the money at stake may differ substantially across contexts.

We could question whether tourists are more likely to "play" with their money than when at home. After dealing with the constant stress of balancing the daily budget at home, tourists may release that responsibility when it comes to vacation time (World of Female 2010). Similarly, tourists' water usage behavior may be different as the entitlement of water at stake varies across locations. For instance, more may be at stake when overconsuming water or not complying with rationing measures at home than when in a foreign location. Notably, if tourists are not as concerned about their spending as when at home, this may weaken the impact of economic incentives when individuals are on vacation.

- *Social norms.* Levitt and List (2007) also suggest that social norms may be triggered differently in context-free lab environments and in context-rich environments. The degree of influence of social norms on compliance behavior will differ across locations. For instance, the "escape from reality" setting that characterizes many vacations is likely to undermine the influence that social norms might exert at home. The social norms surrounding a particular vacation destination may be substantially different from those that exist at the tourist's home location. Tourists may feel compelled to comply with the social norms and expectations of their destination, such as tipping. Alternatively, tourists' behavior may continue to be driven by their local social norms, or these may cease to have an impact on the individual's behavior once he or she leaves home, allowing him or her to feel free, for example, to take longer showers while on vacation. Moreover, the vacation setting lacks the real-life context that may be important in driving certain behaviors at home. Bellis et al. (2004) highlight that vacations provide a diversion from the social mores of family, work, or education. Therefore, social norms may not be influential while an individual is on vacation.

- *Scrutiny.* Levitt and List (2007) also recognize that experimental studies may be less anonymous and therefore subject to an "experimenter demand" effect (Orne 1962). Similarly, an individual's behavior at home is likely to be less anonymous than when on vacation. The literature recognizes that if an individual's compliance behavior stems from moral motivations, then his or her behavior is not as likely to change across situations (see, e.g., Burby and Paterson 1993). Therefore, the source of motivation for sustainable behavior will influence the extent to which scrutiny affects behavior across locations. This suggests that an individual who is motivated by moral drivers is less likely to behave according to the extent of scrutiny. Alternatively, someone who is driven by social norms or the fear of being fined is more likely to behave differently—for example, respond differently to regulations—depending on the degree of scrutiny he or she faces. In addition, moral motivations still have the potential to be undermined; for instance, a care-free environment is likely to challenge moral motivations. Thus, a low degree of scrutiny may lead to socially unacceptable behavior. For example, an individual with high environmental values may use water excessively while on vacation as a result of these factors.

Clearly, there is strong theoretical and practical evidence that vacation and home settings generate different levels of eco-friendly behavior. An individual's eco-friendly behavior might not correlate between settings. If this is the case, eco-friendly behavior in an individual's home city is uninformative about what that individual would do in a vacation setting. If an individual's sustainable behaviors at home and on vacation do not correlate, this can mean either that eco-friendly traits are not stable across situations or that individuals are differently affected by the three factors in Levitt and List's (2007) typology. In the context of this chapter, this also provokes the question whether the same individual's water use behavior at home correlates with his or her behavior while on vacation.

The adoption of eco-friendly behavior may require measures, such as regulations or user-pays pricing structures, to influence individual's behavior. In the context of reducing water consumption in Australia, regulations are the most prominent mechanism (see, e.g., Edwards 2008). For regulations to be effective, individuals need to comply with them, yet motivations to comply differ.

Evidently, an individual's behavior along a number of dimensions differs when he or she is on vacation compared with when at home. Subsequently, an individual may also be motivated to comply with regulations by different drivers in the two settings. Discussion of Levitt and List's (2007) typology indicates that the effectiveness of economic, social, and moral incentives may differ across settings. These types of incentives are explored further in the following section. It is also important to realize that complying with regulations may mean engaging in behavior that is less appealing, less comfortable, or requires additional time for tourists. Although some tourists may be willing to follow regulations, they will invariably need the resources to do so, such as time, money, and information.

TOURISTS AND COMPLIANCE BEHAVIOR

Internal barriers inhibiting sustainable behavior, such as complying with water restrictions, may include a lack of knowledge, the inability to comprehend the consequences of one's actions, and the belief that an individual cannot make a difference (Shove and Warde 2002). The individual's decision to comply is also influenced by external dimensions such as convenience, social norms, and financial resources. The financial implication of a vacation expense for the average household may mean that altruistic arguments advocating for improved attitudes and considerations toward locals and nature may not be effective. The literature suggests that external constraints are more influential than internal knowledge and motivations in shaping tourists' behavior (Kaiser et al. 1999; Tanner et al. 2004). Thus, the types of institutions that govern a society will affect behavior.

Institutions and Compliance Capacity

Institutions can be regarded as "the rules of the game in society or, more formally are the humanly devised constraints that shape human interaction" (North 1990, 3).

Superior institutions can be distinguished by the extent to which the informal "rules of the game," such as in the form of social norms and mores, are consistent with the formal rules developed to govern behavior (Challen 2000; North 1990). This is not to suggest that the informal institutions can substitute for formal institutions in every case (Dovers 2001). However, Dovers (2001) argues that greater alignment of formal institutions with the underlying rules of social networks will give rise to lower costs and therefore superior institutions generally.

This observation has particular relevance in the current context, as understanding of tourists' water-using behavior and preferences is deficient. In most instances, knowledge of an individual's preferences and water use behavior in his or her hometown is not informative of behavior by the same individual as a tourist. Therefore, a city on water restrictions designed to modify the behavior of residents may not have the appropriate incentives or mechanisms in place to drive tourists to comply with water regulations. This has implications for the effectiveness and the cost of policy in this sphere.

Enforcement and the ability to bring compliance to rules has been identified as one of the core features of good institutions generally (North 2000) and for institutions dealing with water allocation and sharing, in particular (Ostrom 1993). There are two basic types of compliance mechanisms: self-enforced and third-party enforced. Notably, different compliance regimes have significant differences in the costs that attend them. The effectiveness of different compliance regimes will also vary across locations and segments of communities. Cooper and Crase (2010) provide empirical evidence of the preferred compliance regimes of water users in particular cities. However, no empirical studies have investigated how an individual's compliance behavior may differ when he or she is recreating or on vacation. This is important for at least two reasons. First, it appears that formal institutions (including those pertaining strictly to compliance) that better align with the underlying motivations of individual behavior will achieve more success and cost less. Second, if tourists self-enforce water restrictions, then water utilities stand to save on the cost of securing compliance. It is against this background that the following discussion is presented.

Theory of Compliance

Enforcement is an important element of regulatory policy design (Cohen 1998) and institutional design generally (Pagan 2009). A trend in regulatory research has been a shift in focus from investigating the enforcement procedures of regulatory bodies to examining the motivations underpinning individual compliance with regulations (see, e.g., Cohen 1998; d'Astous et al. 2005). It seems that the fundamental question underpinning effective development of regulatory policy is why individuals comply with the law. Notwithstanding the research interest in this field (see, e.g., Cooper and Crase 2010), this question is often not considered and applied when stipulating regulatory policy. Those factors that motivate an individual to comply with formal water regulations are of particular focus. An increased understanding of this will assist in identifying what may drive tourists to consume more or less water.

Calculative Motivations. A comprehensive understanding of compliance behavior can support policymakers in developing compliance policy and institutional design of enforcement regimes. Calculated motivations underpin the most established theory regarding regulatory compliance. In Becker's (1968) seminal work, he suggests that the regulated will comply with a particular rule when they perceive the benefits of compliance, including avoidance of fines and penalties, to surpass the associated costs (see also Ehrlich 1972; Stigler 1970). The Beckerian perspective is not without its critics (Wenzel 2005), and a range of alternative motivations are also identified.

Intrinsic Motivations.   A sense of moral obligation (i.e., the need to "do the right thing") is a common driver behind members of society complying, even when the illegal gains exceed the anticipated penalties (Sutinen and Kuperan 1999). Social responsibility or ethics has been identified as a key reason underpinning the adoption of sustainable practices by the tourism sector as "the responsible thing to do" (Tzschentke et al. 2004). Contemporary economics generally fails to recognize that moral aspects have an effect on economic behavior (Hausman and McPherson 1993, 1996).[1] Thus, the adequacy of regulatory policy developed by economists could be challenged.

Moral Development.   Social psychology recognizes the significance of an individual's own characteristics in shaping compliance behavior (see, e.g., Kohlberg 1969, 1984). Researchers propose a positive relationship between the moral development of an individual and his or her propensity to comply with regulations (Sutinen and Kuperan 1999). Kohlberg (1969, 1984) suggests that there are three evident levels of moral development: pre-conventional, conventional, and post-conventional. Pre-conventionalists usually form their rationale on fear of punishment rather than the desire for social order or the potential destructiveness of their behavior; conventionalists generally rationalize on the basis of social conformity and stability; and post-conventionalists are inclined to reason on the basis of moral principles that are independent of social order (Sutinen and Kuperan 1999). Kohlberg (1969, 1984) contends that violation of regulations is likely to diminish at higher levels of moral development, and this has been supported by empirical research in numerous contexts (see, e.g., Kuperan and Sutinen 1999).

Social Motivations.   The concept of social motivation—that is, "the desire of the regulated to earn the approval and respect of significant people with whom they interact" (Winter and May 2001, 678)—has also been recognized as an impetus for compliance. This is consistent with the conventionalist perspective as specified by Kohlberg (1969). The degree to which individuals desire the respect of society is relevant in the context of tourists' water-using behavior. Those who are conscious of how they are perceived by locals will strive to behave consistently with the social norms surrounding water use and regulations to uphold their reputation, notwithstanding that the setting would appear critical. External pressures from consumers and the general community have been recognized as reasons for the adoption of environmental practices in the tourism industry (Middleton and Hawkins 1998).

In sum, both moral obligation and social influence can potentially lead to high levels of compliance, even when a weak deterrent effect exists. In addition to the three fundamental motivations for compliance—calculative, moral, and social— some researchers also factor in the ability and capacity of the regulated to comply (see, e.g., Winter and May 2001).

Ability to Comply.   Willingness to comply is inadequate if individuals do not have the knowledge of what is required of them or are unable to take the requisite steps (Winter and May 2001). It has been suggested that those with a higher awareness of rules will have a greater sense of civic duty to comply, as they are expected to be more cognizant of the reasons for the rules.

Individuals may be unaware of a regulation if it is new, has not been publicized enough, or is too vague or complex for comprehension (Winter and May 2001). Tourists are also likely to be less informed than locals. Borrowing from Carlton and Perloff (1994) and Stiglitz (1989), it is possible to examine markets that comprise both informed customers and uninformed customers. The nomenclature employed for these groups by Carlton and Perloff (1994) is "natives" and "tourists," respectively. One would assume that the natives have a higher level of awareness regarding their city's regulations than do tourists and thus theoretically are more likely to comply.[2] Moreover, the mass media often simply portrays images of destinations rather than providing relevant information for potential travelers. Recognition of the importance of awareness to tourists' responses to regulations (e.g., water restrictions) may improve policymakers' efforts toward increased compliance.

Capacity to Comply.   Awareness and willingness to comply, whether initiated by calculated, normative, or social motivations, are subject to a resource constraint. That is, complying with some regulations may require the existence of financial resources that can be committed to compliance behavior (Winter and May 2001). Compliance may also cost time or be inconvenient, depending on the nature of the rule. For example, water restrictions do not have equal effects on locals and tourists. Tourists, for example, are likely to be unwilling to sacrifice time, such as in the way hand watering is expected of local gardeners.

Potential Complexities.   An additional consideration is that moral drivers could be undermined by economic drivers. More specifically, paying a fine as a result of not complying with a regulation may lead an offender to feel excused for his or her actions, feeling that it justifies the misbehavior and, in effect, buys off his or her guilt (Levitt and Dubner 2005). Thus, an economic incentive may in fact rationalize a breach of a moral incentive and subsequently fail to act as a deterrent at all. Needless to say, behavior is obviously a function of a complex set of variables.

In sum, the literature recognizes a number of variables as playing a part in shaping compliance behavior: severity and certainty of sanctions, potential illegal gain, an individual's standard of personal morality and moral development, and social environmental influences. The concept of compliance is multifaceted, and the literature surrounding this concept is extensive. Tourists' compliance behavior adds another dimension of complexity to this literature.

COMPLIANCE CUBE FRAMEWORK

Cooper and Crase (2010) offer a framework that attempts to capture the pertinent concepts discussed in the compliance literature and assist in understanding their complexity. This framework is presented in Figure 12.1.

Essentially, the compliance cube is used to capture the three key motivational dimensions of compliance: economic, moral, and social (Cooper and Crase 2010). The useful aspect of this framework is that it facilitates the segmentation of individuals according to those motivations that drive them to comply. For instance, individuals that are primarily driven by economic motivations are categorized into the pre-conventional segment; those driven by moral motivations are grouped into the post-conventional segment; and those driven by social motivations fall into the conventional segment.

In addition, the categories are not necessarily mutually exclusive. Thus, individuals who are driven by more than one dimension fall between these extremes. For example, segment *a* in Figure 12.1 would include those individuals driven by both moral and economic motivations.

Based on the moral development literature discussed earlier in the chapter, the framework can be used to identify which segments of the citizenry are most or least likely to comply. The Cooper and Crase (2010) framework is useful for contemplating the challenge of developing effective compliance strategies. For instance, a policymaker that segments a regulated market according to motivations to comply may be able to develop enforcement mechanisms that are more closely

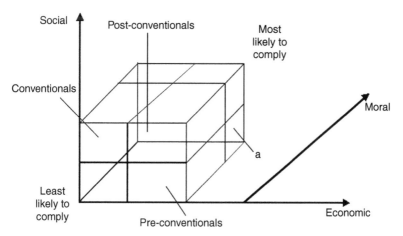

Figure 12.1 *Compliance cube*

Source: Cooper and Crase (2010).

aligned with individual motivations. Therefore, a potentially more cost-effective approach to achieve compliance can be developed.

This framework could be applied specifically in the context of tourists' compliance behavior. There is scope to use the framework to usefully segment locals and tourists and to lower the costs of securing compliance. For instance, if social norms are a prominent dimension of compliance for locals, then it may seem that formal deterrence enforcement mechanisms are unlikely to be cost-effective, at least for that segment. However, tourists may not respond to social norms, perhaps making formal deterrence mechanisms appropriate. Understanding these nuances is critical in designing regulation.

Marketers segment the market to determine which products and services are suitable for particular consumer groups. Similarly, policymakers could segment the market to determine which measures are most appropriate for each group across the population. In the context of water usage, different mechanisms may be required for different segments. In addition, appropriate measures to encourage tourists to adopt eco-friendly behavior may be substantially different from those that are appropriate for locals.

CONCLUSIONS

The literature points to the current and potential problems that arise from tourists' potable water demands. This chapter has attempted to highlight the underlying causes of behavioral differences displayed by individuals in home and vacation settings. In sum, the theoretical and case literature supports the view that tourists are often more prone to abandon the water-conserving behavior employed at home. The policy response for this is a major challenge. Among the most serious considerations is the complexity of relationships and how this impinges on compliance. One option is to critically dissect and segment behavioral motivations. This information could then be mapped to create an improved policy response. Some of these matters are addressed in greater detail in other chapters in this book.

NOTES

1. Different authors have given this dimension of motivation various labels, including "normative commitment" (Burby and Paterson 1993); "moral or ideological compliance" (Levi 1988, 1997; McGraw and Scholz 1991); "commitment based on civic duty" (Scholz and Lubell 1998; Scholz and Pinney 1995); and "an apparent obligation to follow the law," which constitutes a form of legitimacy (Tyler 1990).

2. Conversely, Winter and May (2001) suggest that an increased awareness of rules may foster resistance to them if they are perceived as being irrational.

REFERENCES

Adler, J. 1989. Travel as performed art. *American Journal of Sociology* 94 (6): 1366–1391.
Ajzen, I. 2001. Nature and operation of attitudes. *Annual Review of Psychology* 52: 27–58.

Andereck, K.L. 1995. *Environmental Consequences of Tourism*. General Technical Report INT-GTR-323. Ogden, UT: USDA Forest Service, Intermountain Research Station.

Becker, G. 1968. Crime and punishment: An economic approach. *Journal of Political Economy* 76 (2): 169–217.

Bellis, M.A., K. Hughes, R. Thomson, and A. Bennett. 2004. Sexual behavior of young people in international tourist resorts. *Sexually Transmitted Infections* 80: 43–47.

Bentley, T.A., S.J. Page, and I.S. Laird. 2001. Accidents in the New Zealand adventure tourism industry. *Safety Science* 38 (1): 31–48.

Bergin-Seers, S., and J. Mair. 2008. *Sustainability Practices and Awards and Accreditation Programs in the Tourism Industry: Impacts on Consumer Purchasing Behavior*. Technical Report for the Sustainable Tourism CRC. Queensland, Australia: Griffith University.

Budeanu, A. 2007. Sustainable tourist behavior: A discussion of opportunities for change. *Journal Compilation* 31: 499–508.

Burby, R., and R. Paterson. 1993. Improving compliance with state environmental regulations. *Journal of Policy Analysis and Management* 12: 753–772.

Carlton, D., and J. Perloff. 1994. *Modern Industrial Organization*. New York: Harper Collins College Publishers.

Chafe, Z. 2005. *Consumer Demand and Operator Support for Socially and Environmentally Responsible Tourism*, accessed February 20, 2007, from www.ecotourism.org.

Challen, R. 2000. *Institutions, Transaction Costs and Environmental Policy: Institutional Reform for Water Resources*. Cheltenham, UK: Edward Elgar.

Cohen, E. 1972. Towards a sociology of international tourism. *Social Research* 39: 164–182.

———. 1978. Rethinking the sociology of tourism. *Annals of Tourism Research* 6: 18–35.

Cohen, M. 1998. Monitoring and enforcement of environmental policy in *International Yearbook of Environmental and Resource Economics*, Volume 3, edited by T. Tietenberg and H. Folmer. Cheltenham, UK: Edward Elgar Publishers, 44–106.

Cook, R.L., and K.W. McCleary. 1983. Redefining vacation distances in consumer minds. *Journal of Travel Research* 22: 31–34.

Cooper, B., and L. Crase. 2010. Urban water restrictions: What drives compliance behavior?. paper presented at the 12th Annual Bioecon Conference, September 27-28, 2010, Venice. Venice:.

Cossens, J., and S. Gin. 1994. Tourism and AIDS: The perceived risk of HIV infection on destination choice. *Journal of Travel and Tourism Marketing* 3 (4): 1–20.

Dann, G.M.S. 1983. Toward a social psychological theory of tourisms motivation. *Annals of Tourism Research* 10: 273–276.

d'Astous, A., F. Colbert, and D. Montpetit. 2005. Music piracy on the Web: How effective are anti-piracy arguments? Evidence from the theory of planned behavior. *Journal of Consumer Policy* 28: 289–310.

De Stefano, L. 2004. *Freshwater and Tourism in the Mediterranean*. World Wildlife Fund, accessed August 10, 2010, from www.panda.org/mediterranean.

Dovers, S. 2001. *Institutions for Sustainability, Tela Paper 7: Environment, Economy and Society*, Australian Conservation Foundation. accessed July 13, 2008, from www.acfonline.org.au.

Downing, J., K. Hughes, M. Bellis, A. Calafat, M. Juan, and N. Blay. 2010. Factors associated with risky sexual behavior: A comparison of British, Spanish and German holidaymakers to the Balearics. *European Journal of Public Health* (March): 1–7.

Edwards, G. 2008. Urban water management. In *Water Policy in Australia: The Impact of Change and Uncertainty*. pp. 144–165. Edited by L. Crase. Washington, DC: RFF Press.

Ehrlich, I. 1972. The deterrent effect of criminal law enforcement. *Journal of Legal Studies* 1: 259–276.

Engel, J.F., D.F. Kollat, and R.D. Blackwell. 1968. *Consumer Behavior*. New York, Holt: Rinehart and Winston.

Ewert, A.W. 1994. Playing the edge: Motivation and risk-taking in a high-altitude wildernesslike environment. *Environment and Behavior* 26 (1): 3–22.

Firth, T., and N. Hing. 1999. Backpacker hostels and their guests: Attitudes and behaviors relating to sustainable tourism. *Tourism Management* 20: 251–254.

German Federal Agency for Nature Conservation. 1997. *Biodiversity and Tourism: Conflicts on the World's Seacoasts and Strategies for Their Solution*. Berlin: Springer-Verlag.

Gnoth, J. 1997. Tourism motivation and expectation formation. *Annals of Tourism Research* 24 (2): 283–304.

Gossling, S. 2000. Sustainable tourism development in developing countries: Some aspects of energy use. *Journal of Sustainable Tourism* 8 (5): 410–425.

Gotz, K., W. Loose, M. Schmied, and S. Schubert. 2002. *Mobility Styles in Leisure Time: Reducing the Environmental Impacts of Leisure and Tourism Travel*. Freiburg, Germany: Oko-Institut e.V..

Green, H., C. Hunter, and B. Moore. 1990. Assessing the environmental impact of tourism development: Use of the Delphi technique. *Tourism Management* 11: 111–120.

Hausman, D., and M. McPherson. 1993. Taking ethics seriously: Economics and contemporary moral philosophy. *Journal of Economic Literature* 31 (2): 671–731.

———. 1996. *Economic Analysis and Moral Philosophy*. Cambridge, UK: Cambridge University Press.

Howard, J.A., and J.N. Sheth. 1969. *The Theory of Buyer Behavior*. New York: Wiley.

Honey, M. 2001. Certification programmes in the tourism industry. *UNEP Industry and Environment* 24 (3): 28–29.

Jones, I.C., and J.L. Banner. 1998. Constraining recharge to limestone island aquifers. Geological Society of America 1998 Annual Meeting, *Geological Society of America Abstracts with Programs* 30 (7): 225.

Kaiser, F.G., S. Wolfing, and U. Fuhrer. 1999. Environmental attitude and ecological behavior. *Journal of Environmental Psychology* 19: 1–19.

Knopf, R. 1983. Recreational needs and behavior in natural settings, in *Behavior and the Natural Environment*, edited by I. Altman and J. Wohlwill. New York: Plenum, 205–240.

Kohlberg, L. 1969. Stage and sequences: The cognitive development approach to socialisation, in *Handbook of Socialization Theory and Research*, edited by D. Goslin. New York: Rand McNally, 361–410.

———, 1984. *Essays on Moral Development*, Volume 11, San Francisco: Harper and Row.

Kuperan, K., and J. Sutinen. 1999. Compliance with zoning regulations in Malaysian fisheries. In *Proceedings of the 7th Conference of the International Institute of Fisheries Economics and Trade*. Taiwan.

Levi, M. 1988. *Of Rule and Revenue*. Berkeley: University of California Press.

———. 1997. *Consent, Dissent, and Patriotism*. Cambridge, UK: Cambridge University Press.

Levitt, S., and S. Dubner. 2005. *Freakonomics: A Rogue Economist Explores the Hidden Side of Everything*. New York: Morrow/Harper Collins.

Levitt, S.D., and J.A. List. 2007. Viewpoint: On the generalizability of lab behaviour to the field. *Canadian Journal of Economics* 40 (2): 347–370.

Lewin, K. 1942. *Field Theory of Learning*, Yearbook of National Social Studies of Education. 41: 215–242.

MacCannell, D. 1992. *Empty Meeting Grounds: The Tourist Papers*. New York: Routledge.

Mansfeld, Y. 1992. From motivation to actual travel. *Annals of Tourism Research* 19 (3): 399–419.

Martens, S., and G. Spaargaren. 2005. The politics of sustainable consumption: The case of the Netherlands. *Sustainability: Science, Practice and Policy* 1: 29–42.

McGraw, K., and M. Scholz. 1991. Appeals to civic virtue versus attention to self-interest: Effects on tax compliance. *Law and Society Review* 25: 471–493.

Middleton, V.T., and R. Hawkins. 1998. *Sustainable Tourism: A Marketing Perspective*. Oxford: Butterworth-Heinemann.

Miller, M.L., and J. Auyong. 1991. Coastal zone tourism: A potent force affecting environment and society. *Marine Policy* 3: 75–99.

North, D. 1990. *Institutions, Institutional Change and Economic Performance*. Cambridge, UK: Cambridge University Press.

———. 2000. Understanding institutions, in *Institutions, Contracts and Organizations: Perspectives from New Institutional Economics*, edited by C. Menard. Cheltenham, UK: Edward Elgar, 7–10.

Orne, M. 1962. On the social psychology of the psychological experiment: With particular reference to demand characteristics and their implications. *American Psychologist* 17 (11): 776–783.

Ostrom, E. 1993. Design principles in long-enduring irrigation institutions. *Water Resources Research* 29 (7): 1907–1919.

Pagan, P. 2009. Laws, customs and rules: Identifying the characteristics of successful water institutions in *Reforming Institutions in Water Resources Management: Policy and Performance for Sustainable Development*, edited by L. Crase and V. Gandhi. London: Earthscan, 20–44.

Parrinello, G.L. 1993. Motivation and Anticipation in Post-Industrial Tourism. *Annals of Tourism Research* 20: 233–249.

Pizam, A., G. Jeong, A. Reichel, H. van Boemmel, J.M. Lusson, L. Steynberg, O. State-Costache, S. Volo, C. Kroesbacher, J. Kucerova, and N. Montmany. 2004. The relationship between risk-taking, sensation-seeking, and the tourist behavior of young adults: A cross-cultural study. *Journal of Travel Research* 42: 251–260.

Plog, S.C. 1973. Why destination areas rise and fall in popularity. *Cornell Hotel and Restaurant Administration Quarterly* 14: 13–16.

Rodriguez, A. 1981. Marine and coastal environmental stress in the Wider Caribbean Region. *Ambio* 10: 283–294.

Roehl, W.S., and D.R. Fesenmaier. 1992. Risk perceptions and pleasure travel: An exploratory analysis. *Journal of Travel Research* 30 (4): 17–26.

Scholz, J., and M. Lubell. 1998. Trust and taxpaying: Testing the heuristic approach to collective action. *American Journal of Political Science* 42: 398–417.

Scholz, J., and N. Pinney. 1995. Duty, fear, and tax compliance: The heuristic basis of citizen behavior. *American Journal of Political Science* 39: 490–512.

Shove, E., and A. Warde. 2002. Inconspicuous consumption: The sociology of consumption, lifestyles, and the environment, in *Sociological Theory and the Environment*, edited by R.E. Dunlap, F. Buttel, P. Dickens, and A. Gijswijt. Lanham, MD: Rowman and Littlefield.

Smith, S.L.J. 1990. A test of Plog's allocentric/psychocentric model: Evidence from seven nations. *Journal of Sport Psychology* 4: 246–253.

Stabler, M., and B. Goodall. 1997. Environmental awareness, action and performance in the Guernsey hospitality sector. *Tourism Management* 18 (1): 19–33.

Stigler, G. 1970. The optimum enforcement of laws. *Journal of Political Economy* 70: 526–536.

Stiglitz, J. 1989. Imperfect information in the product market, in *The Handbook of Industrial Organization*, edited by R. Schmalensee and R. Willig. Amsterdam: Elsevier Publishing, 769–847.

Sutinen, J., and K. Kuperan. 1999. A socio-economic theory of regulatory compliance. *International Journal of Social Economics* 26: 174–193.

Tanner, C., F.G. Kaiser, and S. Wolfing Kast. 2004. Contextual conditions of ecological consumerism. *Environment and Behavior* 36: 94–111.

Tyler, T. 1990. *Why People Obey the Law: Procedural Justice, Legitimacy, and Compliance*. London: Yale University Press.

Tzschentke, N., D. Kirk, and P.A. Lynch. 2004. Reasons for going green in serviced accommodation establishments. *International Journal of Contemporary Hospitality Management* 16 (2): 116–124.

Um, S., and J.L. Crompton. 1990. Attitude determinants in tourism destination choice. *Annals of Tourism Research* 17 (3): 432–448.

UN (United Nations). 1995. *Guidebook to Water Resources, Use and Management in Asia and the Pacific*, Volume 1: *Water Resources and Water Use*. Water Resources Series No. 74. New York, United Nations:.

UNCSD (United Nations Commission on Sustainable Development). 2010. *Influencing Consumer Behavior to Promote Sustainable Tourism Development*, accessed August 26, 2010, from http://csdngo.igc.org/tourism/tourdial_cons.htm.

UNESCO (United Nations Educational Scientific and Cultural Organization). 2010. *UNESCO Water Portal Weekly Update No. 155: Water and Tourism*, accessed August 16, 2010, from www.unesco.org.

Upham, P. 2001. A comparison of sustainability theory with UK and European airports policy and practice. *Journal of Environmental Management* 63: 237–248.

VISIT. 2005. *The Tourism Market: Potential Demand for Certified Products*, accessed August 10, 2010, from www.yourvisit.info/brochure/en/070.htm#nachfrage (site now discontinued).

Wearing, S., S. Cynn, J. Ponting, and M. McDonald. 2002. Converting environmental concern into ecotourism purchases: A qualitative evaluation of international backpackers in Australia. *Journal of Ecotourism* 1 (2): 133–148.

Weber, K. 2001. Outdoor adventure tourism: A review of research approaches. *Annals of Tourism Research* 28 (2): 360–377.

Wenzel, M. 2005. Motivation or rationalization? Causal relations between ethics, norms and tax compliance. *Journal of Economic Psychology* 26 (4): 491–510.

Wickens, E. 2002. The sacred and the profane: A tourist typology. *Annals of Tourism Research* 29: 834–851.

Winter, S., and P. May. 2001. Motivation for compliance with environmental regulations. *Journal of Policy Analysis and Management* 20 (4): 675–690.

World of Female. 2010. *How Much Money Do Women Generally Spend on a Holiday?* accessed September 4, 2010, from www.worldoffemale.com.

WTTC (World Travel and Tourism Council). 2007. *The 2007 Travel and Tourism Economic Research*, accessed March 30, 2007, from www.wttc.org.

CHAPTER 13

Water Pricing, Water Restrictions, and Tourism Water Demand

Lin Crase and Bethany Cooper

*I*n Chapter 12, it was argued that the water-using behavior of individuals was likely to be complicated by a variety of factors, including the setting that circumscribed behavior. Specifically in the context of tourism, strong theoretical and empirical grounds exist for suggesting that the use of potable water by individuals might increase when they are tourists or on vacation. Moreover, the factors that shape water-using norms in the domicile setting seem likely to differ markedly from those that operate for tourists and vacationers.

These behavioral considerations have nontrivial policy implications. First, potable water supplies have been critically scarce throughout much of Australia in recent years, and reducing demand to manage shortage has been the core response of most towns and cities. In many instances, this has relied on mandatory restrictions that prohibit or limit specific uses of water, accompanied by vociferous appeals on moral or environmental grounds to encourage modest use of the resource. Second, for locations with a high proportion of tourists or vacationers, the seasonality of tourist demand combined with increasing per capita consumption creates particular water infrastructure challenges. Third, the protracted delays that attend adjustments to water prices render conventional pricing approaches problematic. In simple terms, there is currently a considerable lag between the point at which consumption occurs and any price signal indicating the relative opportunity cost of consumption, especially for tourists. Neoclassical economists would usually advocate for the use of price signals to modify water-using behavior, whether the user is a tourist or local resident. Nevertheless, mounting theoretical and empirical evidence suggests that price signals may not always adequately translate into the behavioral responses necessary to preserve scarce water resources, especially given the way prices are presently established.

This chapter deals with the strengths and weaknesses of alternative water-rationing policies as they relate to tourists and vacationers. It focuses primarily on the water demands of individual tourists, as opposed to the demands on water resources emanating from the maintenance of tourist attractions, such as golf courses and swimming pools. The discussion is also limited to the urban demands of tourists, as opposed to those that arise from nature tourism, which are detailed elsewhere in this volume (see, e.g., Chapter 4).

The chapter first briefly considers the status of urban water restrictions and touches on the efficiency consequences of their imposition. This provides the background for an evaluation of alternative rationing approaches that are applicable to the tourism sector. Following this is an examination of price signals and the intricacies of tariff design. The chapter considers the range of objectives assigned to those who set water tariffs, reflects on the quantity rationing that attends water restriction regimes, and argues for more overt price signals as a vehicle for dealing with water allocation among competing users, including tourists.

WATER RESOURCES, TOURISTS, AND RESTRICTIONS ON POTABLE USE IN AUSTRALIA

The role of water utilities includes providing a degree of supply security to manage the risk of running out of water at some identified or speculated future point in time. This is achieved by water utilities making judgments around supply risks and the trade-offs associated with managing present demands. For instance, trade-offs are made when choosing among harvesting and distributing alternative water sources, augmenting current supplies, and implementing demand-side measures, such as those involved in the enforcement of explicit restrictions on consumption or appeals for water users to limit their use of the resource.

Hypothetically, at least, information operates as a means of moral suasion, influencing personal values, which in turn affects behavior (Fishbein and Ajzen 1975). Politicians favoring water restrictions commonly play the "moral suasion" card (Brennan et al. 2007), and appeals are also made on the basis of intergenerational equity (i.e., use less water to ensure water for your children) (see, e.g., Goulburn Valley Water 2009; Water Corporation 2010).

One of the prominent demand management policies implemented across Australia has been public education and awareness programs promoting water conservation and purported "water use efficiency." For instance, the state government of Victoria has consistently invested in education and awareness campaigns over recent years to encourage water conservation inside and outside the home (see, e.g., Our Water, Our Future 2007; Victorian Government 2003).

In addition, state-owned water utilities are often required to balance a range of broader "social" obligations via their tariffs, such as sending conservation messages, ensuring that the poor are not disadvantaged, and the like (see, e.g., ESC 2009). In balancing these considerations, there is little evidence that water authorities have explicitly considered water consumers' preferences (Hensher et al. 2006), let alone how those preferences might change when an individual moves into vacation mode. Rather, greatest attention is given to the political costs of severe water shortage and the political risks associated with "running dry," and thus the necessity for increasing water supplies.

In contrast to shoring up supplies by tapping new water sources and exerting the moral suasion card, urban water availability can be managed through alternative approaches to demand. One approach involves rationing by either limiting the quantity of water available to customers or invoking some mandatory constraint over water-using behavior. Extensive controversy surrounds the notion of restricting indoor water consumption, primarily because of its intrusive nature, its potential impacts on human health, and the fact that it is associated with "socio-political distaste" (Brennan et al. 2007). In addition, restrictions based on indoor water use face other problems, given the extreme heterogeneity associated with indoor water consumption and the difficulty of monitoring and enforcing indoor water usage. The regulation of outdoor water consumption is not only more socially acceptable, but also more plausible. Regulations can be implemented by means of restricting usage to particular days of the week, times of day, and forms of watering devices (e.g., hand-held hoses instead of sprinklers). More important, given that outdoor water use can be classified as "conspicuous consumption," these restrictions can potentially be enforced through other community members reporting their neighbors for recalcitrant behavior. Regardless of the practicality of this approach, it is not without its costs, as discussed later in this chapter.

Changing rainfall patterns and lower streamflows across Australia have led to a focus on a range of conservation measures and "demand management" initiatives. Based on the rationale above, these mainly target outdoor use, which is perceived as "discretionary" (NWC 2008). State governments have responded to the problem in a variety of ways. Most began by implementing progressively more stringent conditions on household water use, while simultaneously (and strangely) mounting arguments against augmentation of water storage infrastructure such as dams and weirs (Byrnes et al. 2006). By May 2007, five capital cities around Australia had resorted to water restrictions to manage diminishing supplies (NWC 2007).

Several states in Australia currently impose urban water restrictions on a temporary basis, while others have moved to place them on a permanent footing. For instance, cleaning paved surfaces by hose is prohibited in Victoria, regardless of the volume of water on hand. Restriction regimes that include sets of rules, stages, triggers, and levels of service vary considerably across Australia.

Important efficiency implications attend a restrictive regime of this form. Some of these stem from the substantial inconvenience and welfare losses imposed on urban water users. In this context, several empirical studies have been undertaken to give some indication of the costs imposed by restricting the use of potable water. Different approaches have been employed in this regard. For example, Grafton and Ward (2008) estimate the welfare costs of water restrictions by comparing the change in surplus that would result from removing all water restrictions in Sydney and replacing them with a water price of A\$2.35/kiloliter.[1] Using 2004–2005 data, they estimate that the costs of restrictions in that year were about A\$55 per person or A\$150 per year per household. Using a different approach, Brennan et al. (2007) estimate the welfare losses associated with different levels of sprinkler bans in Western Australia. In this instance, welfare losses are estimated by manipulating the production function for lawns and establishing the labor substitution requirements at different labor rates. When outdoor watering of lawns is restricted to two days per week, the cost to households is in the order of A$100 for the season. A total sprinkler ban sees these costs rise to between A$347 and A$870, depending on the opportunity cost of time for each household, with a typical household cost of about A$487.

An important observation from the current rationing mechanisms is that they almost entirely relate to the water-using behavior of local residents. For instance, it is difficult to imagine the circumstances under which tourists or vacationers might be directly affected by a sprinkler ban at particular times of the day. Moreover, because indoor water use is not subject to constraints, there is a risk that rule-based behavioral restrictions will do a poor job of moderating excessive water use by vacationers or tourists. Put differently, water restrictions are not only costly for urban water users generally, but they are also inequitable and have more substantial impacts on local residents than on tourists or vacationers. On these grounds alone, it would seem sensible to consider alternative rationing measures, especially in locations where tourists make up a large portion of potable water users.

INFLUENCING TOURISTS' WATER CONSUMPTION

Setting aside the demands that arise from tourist attractions, water use by tourists on a per capita basis has been shown to far exceed that of local residents (see, e.g., De Stefano 2004; Narasaiah 2005). Tourists staying in hotels commonly use 30% more water than others in the same location (EEA 2003). At a national level, the empirical research in this context is patchy, at best. Studies that specifically address the motivations of individual tourist water use and measures that might be used to rein in consumption in times of scarcity are even rarer.

As noted earlier, the rule-based regimes that are presently in place ostensibly have no direct impact on tourists, especially in the context of indoor water use. Yet indoor water use commonly accounts for about 60% of all potable water consumption. Moreover, if tourists constitute a substantive portion of the population of a city, and their water-using behavior is more profligate than the norm, the demands on water resources and related infrastructure are likely to be significant. Thus, modifying demand requires targeted policy measures.

Against this background, concern that tourists are overconsuming water has led to questions regarding the awareness levels of this cohort (see, e.g., Westernport Water 2010a). Subsequently, a number of initiatives have been targeted at tourists' water-using behavior, attempting to convey the message that excessive water use is unacceptable in *all* settings. For instance, Westernport, Wannon Water, and Barwon Water, which service Victorian coastal vacation destinations, support a campaign designed to remind visitors not to take a "holiday" (the Australian term for vacation) from saving water. Essentially, "caravan" (recreational vehicle) parks, motels, restaurants, and fast-food outlets are targeted with posters developed to remind visitors along Victoria's coast about the "precious" water supplies in these destinations (Westernport Water 2010a). Although such campaigns might demonstrate to local residents that the water utility is treating their concerns seriously, the effectiveness of these measures on tourists' and vacationers' behavior is debatable.

Tourism Australia (2010) encourages accommodation providers to place notices about water conservation that identify ways staff and guests can minimize individual water consumption. For instance, reminding guests to take short showers is a common tactic. The average medium to large hotel in Australia uses about 79,000 liters per day (or 301 liters per room per day), with showering being the main water-using activity. It has been argued that consumption could be reduced by not only educating guests, but also giving tourists the option to time their showers (Ecogreen Hotel 2008). A shower timer is just one of the many "educational" products designed to align behavior with conservation messages (see, e.g., Save Water 2005). Moreover, linen and towel reuse programs can considerably reduce the amount of water used in the laundry. This type of program has become increasingly popular in hotels, largely owing to the positive reactions of guests (Project Planet 2010).

Notwithstanding the benefits of this approach, an important cautionary note applies. Notably, Cooper and Crase (2009) point out that the cost of educational and awareness campaigns versus the wider gains is seldom given serious consideration. Moreover, their research finds that preferences for informational campaigns are hardly homogenous. More specifically, they find that in some instances, people will pay extra to avoid continuous exposure to what might be regarded as crass environmental fundamentalism or never-ending appeals to conserve water.

A number of physical devices can be installed in tourist accommodation facilities to assist in reducing water use by guests. Tourism Australia (2010) encourages the adoption of several measures. First, it recommends installing highly rated water-efficient showerheads, which use about 10 liters of water per minute, whereas older models use 20 to 30 liters per minute. Moreover, using water-efficient showerheads decreases the amount of hot water needed and thus reduces the energy requirements of the facility. Second, it suggests using passive infrared sensors for urinals in men's restrooms. These normally use 20% less water than a conventional urinal flushing system. Third, it encourages installing dual-flush or water-saving toilets, which use 4 to 6 liters per flush, whereas a standard single-flush toilet uses 10 to 15 liters. Tourism Australia (2010) points out that, looking at the larger picture, in a 100-bed hotel, dual-flush toilets could save over 200,000 liters of water per year. Notably, water-efficiency audits in the Australian hotel industry indicate that water use can be decreased by an average of 20% without compromising guest comfort (Tourism Australia 2010). This does not imply, however, that such measures are cost-effective from the perspective of the accommodation proprietor or even socially desirable. Nevertheless, it gives some indication of the lengths to which some providers will go to constrain water consumption demand.

At the heart of this issue are two core problems relating to the incentives given to and signals received by individual tourists. First, tourists by definition are temporary residents at any destination. In this context, the costs that arise from excessive water consumption take on the form of an externality insomuch as tourists may simply go to another location when diminished supplies manifest in inconvenience. Accompanied by the fact that water restrictions focus on indoor (discrete) water use, the individual incentives for conservation are weak. Second, costs do not directly flow to the present generation of tourists. For example, if additional water supply augmentation is required because of excessive water demand, the costs of augmentation works will ultimately be factored into water tariffs. Depending on the degree of competition in markets, one might expect that all or some of these costs ultimately result in increased tariffs in hotels or other accommodation outlets. However, this will have direct impacts on tourists or vacationers only if they make return visits to the destination or choose to stay for an extended period of time. Again, the capacity of price to send an appropriate signal is muted by the chain of events to realize a modification to price. In this context, it is worth briefly reflecting on the mechanics of establishing water tariffs in Australia.

WATER TARIFF DESIGN AND TOURISTS' USE OF POTABLE WATER

Urban water tariffs have undergone substantial reform since the ratification of the National Water Initiative (NWI) by all Australian jurisdictions in 2004. An important ingredient of the NWI was a call for the implementation of "best practice water pricing" and "improved pricing for Metropolitan water." However, considerable discrepancy exists around what constitutes "best practice" in this setting (see, e.g., Dwyer 2006). In the current context, two elements warrant particular attention: the time frame for adjustments to water tariffs and the relative weighting of fixed and variable components of water tariffs.

Adjusting Water Tariffs to Reflect the Cost of Use

Water and wastewater services are generally regarded as monopolies in Australia. The public policy response has been to retain water and wastewater services in public hands and to have prices be subject to some form of economic regulation. The institutional model for water and wastewater services varies considerably across and within state jurisdictions. For example, water and wastewater services in Victoria are provided by state-owned water corporations with geographic monopolies over water supply districts. Prices must be approved by the Essential Services Commission, which endeavors to match a utility's revenue with the marginal cost of provision. By way of contrast, in New South Wales, metropolitan water and wastewater services are provided by Sydney Water, a state-owned entity, but nonmetropolitan services are provided by local governments. In the case of the former, economic regulation is mandatory via the New South Wales Independent Pricing and Review Tribunal (IPART), whereas most nonmetropolitan water prices must meet guidelines set down by the state government. The other states have different configurations of service delivery and regulation, although government ownership is the norm.

One common feature of each jurisdiction is the capacity to vary prices in the short and medium term. Water utilities are generally required to submit medium-term (say five-year) projections of water demand and details of core supply augmentation works to maintain some minimum but acceptable level of service. This then becomes the basis on which water tariffs are enumerated, or at least the basis for the revenue collected by the utility in that period. Clearly, there is considerable scope for strategic behavior in this setting, with utilities invariably opting to overestimate the growth of demand, the necessity for supply augmentation, and the level of service that might be regarded as "acceptable." In contrast, the economic regulator will usually seek to deflate the growth of demand; limit capital expenses for augmentation works, especially if they are considered "gold plating;" and challenge the minimum service requirements. In any case, a core consideration is the time lag between these negotiations and a modification in water prices to reflect any changed circumstances.

In the case of an ongoing shortage of water, the utility will first notice a decline in reliability and invariably a fall in revenues, especially from the volumetric contribution of tariffs, discussed later in this chapter. The upshot will be that the utility is obliged to consider augmentation of water supply. Depending on the geographic and hydrologic setting, this could take the form of developing new water-harvesting structures, tapping groundwater supplies, purchasing additional water rights from other users, or sophisticated engineering works such as recycled wastewater substitution and desalination, to name a few. Ideally, each option would be subject to some consideration in line with its order of economic merit. In reality, a host of political influences may come to bear, say in the form of resistance to water trade away from agriculture to urban users or excessive political enthusiasm for recycling on perceived environmental grounds (see, e.g., Crase et al. 2007).

In any case, the supply augmentations will be subjected to the scrutiny of the economic regulator and, following some negotiation, ultimately will be factored into the required capital expenditures of the utility in the subsequent planning period. This will then manifest in increased tariffs faced by customers to cover the economic costs of the supply augmentation.

Tortuous debates can inevitably circumscribe decisions of this form. Some pertain to the merits of one supply option over another. Other controversies will be raised about more detailed elements, such as the appropriate rate of return on these investments and the extent to which they should be treated as sunk. Other questions will undoubtedly be raised about the extent to which returns to publicly owned entities should mirror other capital markets. From the point of view of the current discussion, there is one clear element of concern—the delay between excessive consumption and the resultant price signal to water users.

Partially stimulated by these concerns, Grafton and Ward (2008) offer the notion of a scarcity price as a vehicle for managing excessive water demand. In essence, they argue that water prices should rise as current available supplies diminish, thereby instantaneously choking off demand. Conversely, in times of plenty, water prices should fall, again reflecting the scarcity value of the resource. In essence, this is what currently happens with water prices, although the regulatory framework is sufficiently cumbersome to mute all but the loudest price signals. Water prices do rise in response to scarcity, but only after the exhaustive processes of assessing the cost of supply augmentation are completed and this is modeled into tariffs. Clearly, these arrangements are exacerbated by tourists' behavior, especially if the price signal arrives belatedly and the water user is already gone. In essence, the current delays in adjusting water tariffs make it difficult for domicile water users to respond to price signals, let alone tourists, who face a derived and delayed price signal at best.

Fixed versus Variable Water Tariffs

In general terms, most of the regulatory activity around water tariffs centers on the revenue required by monopoly suppliers to continue adequate service provision. Much theoretical literature has been devoted to how this revenue should be collected. More specifically, considerable debate has surrounded the merits of focusing on short-run marginal cost (SRMC) and long-run marginal cost (LRMC). SRMC refers to the cost of meeting an additional unit increase in demand within an existing supply system. If the system has excess capacity, this will result in a low revenue requirement and low water price. LRMC is generally favored by economic regulators in Australia and is implicit in the NWI (Edwards 2007). This approach considers the cost of augmenting supply infrastructure as capacity is progressively exhausted. Debate also pertains to whether cost recovery is best achieved by modifying fixed or volumetric water charges.

From a theoretical perspective, fixed charges or access charges should be set in a manner that recovers the fixed costs associated with the provision of water services. These arrangements will purportedly lead water users to consider all alternative investment options and simultaneously encourage water utilities to undertake only those supply augmentations that can pass economic muster. In reality, neither of these arguments tends to hold (Crase et al. 2008), and regulators resort to more pragmatic arguments for apportioning the total revenue required by the utility to either fixed or volumetric categories.

One of the consequences of imposing higher volumetric charges is that the price signal received by water users is more overt. Significantly, research by Crase et al. (2008) showed that the majority of urban water users actually preferred a tariff structure weighted toward volumetric charges, even where the household was a large water user. However, fixed charges ensure the utility of revenue regardless of whether consumers use more or less of the resource. Another implication of a disproportionate reliance on fixed charges is that the water utility's incentive for finding innovative means to meet demand is weakened. Where the utility is reliant on selling water for its income, it is more inclined to find alternative supply sources. In this context, Crase (2009) found that water utilities located in neighboring cities displayed very different market responses: the utility with a high fixed charge was inclined to impose rule-based restrictions on water use, whereas the utility with higher volumetric charges willingly accessed the water market to limit the impact of behavioral constraints.

In destinations with a highly seasonal tourist demand, the split between fixed and variable charges has important nuances. In many tourist destinations, water (and wastewater) infrastructure must be constructed to deal with peak loads. For example, the Victoria surf coast is serviced by Westernport Water, which recently reported that its 17,000–customer base swells to 60,000 over the summer months (Westernport Water 2010b). The upshot is that potable water treatment plants, clear-water storage structures, and the like must all carry considerable excess capacity for nine months of the year. Efficiently distributing the costs of this capacity is problematic.

On the one hand, the water utility could opt for a seasonal (peak-load) tariff so that beneficiaries become obliged to face the marginal cost of their tourism choices. Moreover, in the case of Westernport Water, this is a preferred approach, although it is not without its flaws. For instance, the Essential Services Commission notes that although aiming to target summer usage, the protracted billing cycle used by the water utility would ostensibly amount to an eight-month peak price. This could well be construed as a winter discount rather than a summer peak pricing structure, "further weakening long-term conservation signals" (ESC 2005 176). This also says nothing of the considerable revenue exposure of the utility, as a downturn in tourism due to a range of exogenous factors might threaten the capacity of the utility to recoup costs.

On the other hand, the problems of peak-load infrastructure and cost recovery could be dealt with by imposing a flat usage charge accompanied by a minimum usage bill. This was recommended by the Essential Services Commission on the grounds that it would enable the water utility to allocate the additional costs associated with peak demand to nonpermanent residents via the minimum usage charge (ESC 2005). In effect, these arrangements tend to increase the fixed charge for nonpermanent residents. Not surprisingly, they limit the extent to which usage at the margin is confronted by cost and, perhaps ironically, result in the perverse incentive for tourists to use more water during the holiday period.

Similar problems beset the commonly employed inclining block tariff (IBT) structures that are widely deployed in the urban water sector. IBTs comprise several volumetric tariffs, which, as the name implies, rise as the volume of water consumed increases. The rationale for this approach is that a modest quantity of water should be provided to meet essential human needs and that this has public-good attributes, such as in the form of guaranteeing hygiene and health standards. Thereafter, water consumption is deemed "discretionary" and should thus philosophically face a higher price. However, although they are politically appealing, IBTs have been found wanting on many fronts. For instance, the point at which water consumption becomes discretionary is subjective and subject to political manipulation (see, e.g., Crase et al. 2007). At a practical level, the price of the lower-bound block must be discounted to prevent the utility from recovering more than the minimum revenue to justify provision (Brennan et al. 2007). The upshot is that because water consumption is heterogeneous, the risk is great that wealthy low-volume water users will ultimately be subsidized by poorer high-volume water users. In the context of tourism and peak demand, an IBT accentuates this potential cross subsidy. For instance, if wealthy vacationers occupy a vacation residence for a few months of the year, it is unlikely that their water consumption will exceed the lower-bound pricing blocks, resulting in the volumetric charge they face being less than that faced by others. Clearly, designing a tariff structure in which tourists face the marginal cost of their water use is not likely in the near future, especially with existing technologies, billing cycles, and institutional arrangements.

Recent attempts have been made to provide immediate signals to tourists regarding their water consumption. For instance, smart meters offer a relatively immediate signal to tourists regarding their water consumption. Smart meters have been used in residential households and by tourism operators (Our Water, Our Future 2010). Smart meters are devices that have the capability of providing tourists with instant (or "real-time") volumetric monitoring and continuous water use data over the period of their stay. This assists with improving tourists' understanding of water use. To date, such devices have been used primarily as an educational or assuasive tool rather than as a vehicle for directly dealing with the peak-load pricing issues associated with water infrastructure in tourism destinations (see, e.g., Our Water, Our Future 2010). This is not to say that such approaches do not warrant further investigation, especially in the context of the already considerable welfare losses associated with inefficient pricing structures and behavioral restrictions.

CONCLUSIONS

Tourists have historically used a much greater quantity of potable water than do local residents. This is a significant issue because it leads to major infrastructure challenges and threatens the long-term assurance of water availability in many destinations. In the urban water setting in Australia, the convention has been to invoke a range of rule-based behavioral restrictions that attempt to limit demand. The welfare costs associated with these approaches are gaining increased attention in the literature, and they are now widely regarded as an unsustainable long-term policy response. Violations of equity also occur, insomuch as tourists are largely unaffected by mandatory water restrictions.

Given the limitations of rule-based water restrictions, water utilities and governments have largely relied on educational and informational campaigns to encourage conservation behavior by tourists. Tourism and accommodation operators have also endeavored to rein in water use through the installation of "water-wise" devices and placement of messages to remind tourists about the wider merits of water conservation. In the light of the literature reviewed in Chapter 12 and other concerns about the effectiveness of such approaches, greater attention to price signals appears warranted.

Nevertheless, this is no straightforward matter. Water tariffs must invariably meet a suite of criteria beyond their conservation objective. In this context, there are major limitations to the way tariffs are developed, implemented, and reviewed.

Invariably, tourists face relatively weak price signals about their water-consuming behavior. Some technologies are emerging that could improve the effectiveness of price as a rationing device, and these appear to offer promise. Ultimately, the merits of different rationing approaches will change over time, and institutional impediments to more efficient pricing should be challenged.

NOTE

1. A\$1 = US\$0.9978 as of January 2011.

REFERENCES

Brennan, D., S. Tapusuwan, and G. Ingram. 2007. The welfare costs of urban outdoor water restrictions. *Australian Journal of Agriculture and Resource Economics* 51 (3): 243–261.

Byrnes, J., L. Crase, and B. Dollery. 2006. Regulation versus pricing in urban water policy: The case of the Australian National Water Initiative. *Australian Journal of Agricultural and Resource Economics* 50 (3): 437–449.

Cooper, B., and L. Crase. 2009. Urban water restrictions: Unbundling motivations, compliance and policy viability, paper presented at the *Australian Agricultural and Resource Economics Society 53rd Annual Conference*, February 10–13, 2009, North Queensland.

Crase, L. 2009. Water—the role of markets and rural-to-urban water trade: Some observations for economic regulators. *Connections: Farm, Food and Resources Issues* 9: 1–5.

Crase, L., S. O'Keefe, and J. Burston. 2007. Inclining block tariffs for urban water. *Agenda* 14 (1): 69–80.

Crase, L., S. O'Keefe, and B. Dollery. 2008. Urban water and wastewater pricing: Practical perspectives and customer preferences. *Economic Papers* 27 (2): 194–206.

De Stefano, L. 2004. *Freshwater and Tourism in the Mediterranean.* Rome: WWF Mediterranean Programme.

Dwyer, T. 2006. Urban water policy: In need of economics. *Agenda* 13 (1): 3–16.

Ecogreen Hotel. 2008. *Water,* accessed August 16, 2010, from www.ecogreenhotel.com.

Edwards, G. 2007. Urban water management, in *Water Policy in Australia: The Impact of Change and Uncertainty,* edited by L. Crase. Washington, DC: RFF Press, 144–165.

EEA (European Environment Agency). 2003. *Europe's Water: An Indicator-Based Assessment,* accessed August 30, 2010, from www.eea.europa.eu.

ESC (Essential Services Commission). 2005. *Water Price Review,* Volume 1: *Metropolitan and Regional Businesses' Water Plans, 2005–06 to 2007–08.* Melbourne: Essential Services Commission.

———. 2009. *Metropolitan Melbourne Water Price Review 2008–09: Final Decision.* Melbourne: Essential Services Commission.

Fishbein, M.A., and I. Ajzen. 1975. *Belief, Attitude, Intention and Behavior: An Introduction to Theory and Research.* Reading, MA: Addison-Wesley.

Goulburn Valley Water. 2010. *Saving Water,* accessed January 10, 2009, from www.gvwater.vic. gov.au.

Grafton, Q., and M. Ward. 2008. Prices versus rationing: Marshallian surplus and mandatory water restrictions. *Economic Record* 84: 57–65.

Hensher, D., N. Shore, and K. Train. 2006. Water supply security and willingness to pay to avoid drought restrictions. *Economics Record* 256 (82): 56–66.

Narasaiah, M.I. 2005. *Water and Sustainable Tourism.* New Delhi: Discovery Publishing House.

NWC (National Water Commission). 2007. *National Performance Report, 2005–2006: Major Urban Water Utilities.* Melbourne: Water Services Association of Australia.

————. 2008. *National Performance Report, 2007–2008: Urban Water Utilities*. Melbourne: Water Services Association of Australia.

Our Water, Our Future. 2007. *Using and Saving Water*, accessed November 10, 2009, from www. ourwater.vic.gov.au.

————. 2010. *Smart Water Metering Cost Benefit Study*, accessed August 19, 2010, from www. ourwater.vic.gov.au.

Project Planet. 2010. *Guest Responses*, accessed August 30, 2010, from www.projectplanetcorp. com.

Save Water. 2005. *Water Efficient Products*. accessed August 16, 2010, from www.savewater.com. au.

Tourism Australia. 2010. *Water*, accessed August 16, 2010, from www.tourism.australia.com.

Victorian Government. 2003. *New Campaign Urges Melburnians to Be Water Savers*. media release from the Office of the Premier, accessed March 10, 2009, from www.legislation.vic.gov.au.

Water Corporation. 2010. *Water Mark: Water for All, Forever*, accessed January 8, 2010, from www.watercorporation.com.au.

Westernport Water. 2010a. *Revisiting the Holiday Water-Saving Message*, accessed August 19, 2010, from www.westernportwater.com.au.

————. 2010b. *Westernport Water Annual Report*, accessed October 24, 2010, from www. westernportwater.com.au.

CHAPTER 14

Lessons for the Tourism and Recreation Sector and Directions for Future Research

Sue O'Keefe and Lin Crase

T he aim in bringing together a range of contributions addressing the relationship between water and the tourism and recreation sector in Australia was to delve into some of the complexities that attend this relationship and thereby reduce the existing research void. In designing this book, we sought to divide the discussion into four parts that would present coherent yet related arguments. Each part focused on one of the four central questions raised in Chapter 1. Contributors have raised a range of important public policy and institutional issues with lessons for a wider and geographically dispersed audience. This chapter summarizes the salient findings of the previous chapters as they relate to the four central questions and extends this discussion to examine areas for future research. The chapter is organized around the four key questions.

PRODUCTION RELATIONSHIPS, VALUES, AND TRADE-OFFS

The first key question is, what is known about the value of water for tourism and recreation and, more specifically, about the conflicts and complementarities among the various uses of the resource? Part I focused on matters relating to production relationships, values, and trade-offs, as these underpin much of the following discussion. As a backdrop to the ensuing chapters, Chapter 2, by Simon Hone, highlighted the importance of environmental and scientific considerations in current water policy debate. In the context of tourism and recreation, there are both complementarities and conflicts with environmental claims on the water resource. Moreover, an examination of the science of Australia's rivers provided a necessary context to the work of other contributors. In this chapter, the perilous state of Australia's waterways was chronicled, with particular emphasis on the effects of human intervention. Hone examined some of the complex linkages among hydrology, water quality, and habitat, using a simplified production systems perspective to conceptualize relationships between inputs and outputs. This chapter included practical examples of the approaches to the evaluation of ecosystem health and the restoration of the Snowy River. Hone's contribution also underscored the complexities involved in ecological health. This has particular resonance in Australia, where environmental flows are conceptualized almost exclusively in terms of volume. Hone's analysis showed that the delicate ecological balance necessary to achieve healthy rivers and ecosystems is a function not only of volume, but also of frequency, duration, timing, and rates of change. This has important implications for the design of water rights—a matter taken up in detail in Chapter 5.

Any allocation decision must take into consideration the value of the resource in its various uses, and this has proved to be a particularly vexatious issue in the context of tourism and recreation. Because the value of water for tourism transcends use values and takes place outside the traditional market, calculation is fraught with difficulties. Nonetheless, without a real understanding of the value of water in this context, it is likely that allocation decisions will be suboptimal. In Chapter 3, Darla Hatton MacDonald, Sorada Tapsuwan, Sabine Albouy, and Audrey Rimbaud surveyed the existing literature and reported on the range of studies employed to value tourism and recreation in the Murray-Darling Basin. Importantly, they argued that an understanding of value is crucial to the formulation of appropriate policy. Sensible water policy should take account of the value of water in various contexts, including nonmarket values. It should also take account of the fact that values change over time and the necessity to limit institutional impediments that would prevent water from moving to reflect these changes.

Hone's contribution also set the scene for the detailed examination of trade-offs, a theme taken up in the context of ecosystem services in Chapter 4. Here, Pierre Horwitz and May Carter explored the links among tourism, ecosystem services, human well-being, industry setting, and visitor experience, and proposed a way to improve the management of inland freshwater systems. Central to their argument are the notions that the management of water resources for tourism and recreational use is accompanied by important trade-offs in ecosystem services, and that compromise and trade-offs are inevitable. They developed a novel framework for the conceptualization of trade-offs in terms of ecosystem services and contended that successful management regimes must recognize and negotiate these compromises. This is likely to be no easy task, as illustrated by political events in Australia around the implementation of a Murray-Darling Basin Plan.

The framework developed in this chapter built on the concepts of value, conflicts, and complementarities and offered a way forward in the management of fresh water for multiple uses. Importantly, it included both social and cultural values and conceptualized the notion of trade-offs through the language of ecosystem services. The great benefit of this approach is that it allows trade-offs to be made explicit at any stage of the planning process and acknowledged by stakeholders. Thus, having identified the likely trade-offs associated with a given visitor experience, the question of compensation can be addressed through both market and nonmarket instruments. For example, payment for environmental services can be used to acknowledge the consequences of particular management decisions. The success of this approach relies on the inclusion of the public, politicians, economic interests, and regulators in the process. This inclusiveness is a matter taken up in depth in the theoretical arguments presented in Chapter 11.

Collectively, the first four chapters contribute much to our way of conceptualizing value, conflicts and complementarities, and the associated notion of trade-offs. They also point the way for further examination on a number of fronts. The key lessons derived from this section relate to the complexities of valuing water for tourism and recreation and the difficulties that arise in quantifying trade-offs, especially given that current policy conceptualizes water only in terms of volume. The institutional arrangements surrounding the management of water for tourism and recreation were addressed in Part II.

PROPERTY RIGHTS, INSTITUTIONS, AND DECISIONMAKING

The second question is, how does the crafting of property rights and related institutions constrain or influence tourism water management decisions? Part II of this book raised some important institutional lessons relating to the property rights to water and the potential of collaborative and devolved models of water management that can have impacts on tourism interests. In Chapter 5, Lin Crase and Ben Gawne argued that the crafting of property rights is no straightforward matter, and that the "shape" of rights affects the behaviors of various players within the water sector. Expanding on the point raised in Chapter 2, Crase and Gawne noted that rights have more than one dimension. Water is fugitive, and consumption at one point or at one time does not of itself prevent further consumption at other points in space and time. A range of incidental or nonconsumptive benefits (and costs) also attend water simply by virtue of its presence at a point in space or time. Thus, specifying exclusivity, access, management, and the like becomes problematic for water resources. Dimensions of volume, variability, and timing all have important ecological impacts (a point raised by Hone) in addition to substantial effects on the tourism and recreation sector. Crase and Gawne argued that the current specification of rights in volumetric terms is inadequate for the needs of the tourism and recreation sector and suggested an approach that may better suit this sector. More specifically, unbundling rights such that delivery times and the management of water storage become more contested offers considerable promise. Without such rights, interests such as tourism and recreation will have few options other than influencing water allocation via the political process.

In Chapter 6, Brian Dollery and Sue O'Keefe argued that Australian water policymaking has highlighted the systematic weaknesses of "top-down" decisionmaking. They examined the literature of public choice and referred to a new paradigm that has at its core broad-based participation, local knowledge, and a collaborative approach to dealing with the inherent uncertainty embodied in natural resource management. They drew attention to Ostrom's (1990) principles for the design of decentralized solutions, including clearly defined boundaries for participation, the adaption of rules to local conditions, participation of users and potential users, effective monitoring, and recognition of the self-determination of the local community by higher-level governing bodies. However, they noted that despite their potential, collaborative models like the hybrid partnership model should not be seen as a panacea for debates on environmental and tourism policy that are inherently complex and political in nature.

The theoretical examination of these complex issues was followed in Chapter 7 with Sue O'Keefe and Brian Dollery examining the potential for water trusts to bring together environmental and tourism and recreational interests. This would ostensibly entail entering the water market to secure allocation outcomes that coincide with particular interests. Water trusts have had success in the United States, and they appear to offer several advantages over other decisionmaking institutions. Central to these is their potential to develop innovative solutions to allocation problems and build relationships by negotiating with irrigators at the local level. O'Keefe and Dollery examined the operation of water trusts in the western United States and questioned whether similar institutional arrangements could provide for enhanced outcomes in the Australian context. They argued that despite some embryonic activity of this kind in relation to environmental interests, several questions remain unanswered. Foremost among these is the complementarity between the water needs of the environment and the tourism and recreation sector and the problems associated with the specification of rights in purely volumetric terms—matters that were taken up in other chapters in this book.

In sum, these chapters clearly illustrated the ways in which property rights and other institutions constrain the allocation decisions of water managers. The various dimensions of water, and aspects of its value in both use and nonuse senses, mean that there is scope for an unbundling of water property rights to go beyond the volumetric conceptualization currently in use. This will require harnessing information about production relationships and attendant trade-offs to determine the scope for a portfolio of rights that more accurately reflect the needs of both the environment and the tourism and recreation sector. Lessons can be learned from the experience of water trusts in the United States, which have developed "products," such as split season leases, to take into account (and to value) the timing of irrigation flows. The overall merits of these changes constitute an empirical question that requires further investigation, however, and it is unlikely that the US experience can be simply transplanted onto an Australian setting. Nonetheless, these strands of inquiry are likely to yield improved outcomes for the tourism and recreation sector, as well as increase their broader appeal from a policy perspective.

CURRENT ISSUES IN WATER POLICY FOR TOURISM AND RECREATION

The third question is, what does current practice tell us about policy formulation, and how can knowledge be better developed and aligned to serve the interests of competing users? Part III of this volume presented examples of current practice and particular challenges for the tourism sector. The aim was to derive lessons that might enable the sector to better influence policy. Cases were scrutinized in an attempt to understand how tourism and recreation interests might set about harnessing important complementarities with other users. More specifically, the chapters in this section provided an opportunity to consider the nexus between the knowledge of water and the politics of water. This approach offered an alternative lens to conceptualize how water policy is formulated that derives from a set of significant politicoscientific lessons. This section comprised three chapters that served as case examples, followed by a broader contribution that drew the previous chapters together.

In Chapter 8, Fiona Haslam McKenzie highlighted the curious lack of activity and development on the beautiful Swan River in Perth and contrasted this with the development of a vibrant river precinct in Harlem, New York. She described a range of political, social, and institutional factors that appear to have prevented the further development of this natural West Australian asset. Although the Swan River and Harlem River have had similar experiences, with an area of contested real estate and environmental damage, Harlem was able to overcome these impediments. This provides several salient lessons that are echoed throughout this book. First, it is essential to balance the interests of the relevant stakeholders to avoid suspicion and mistrust. Second, the way to achieve this is through engagement of the community in the planning process. And third, good governance is the key to achieving outcomes that are acknowledged as benefiting all.

Michael Hughes and Colin Ingram investigated in Chapter 9 the practice of exclusion from urban water catchments in Perth and noted that in this respect Perth appears to be out of step with the approach taken elsewhere in the world. Exclusion carries with it costs that relate to human health and well-being, and the authors investigated alternative institutional structures that would appear to offer an enhanced result for those seeking a tourist or recreation experience. Hughes and Ingram advocated strongly for a more closely integrated management regime based on a comprehensive understanding of values, including social, recreational, and tourism values, alongside the value of high-quality potable water. Thus, when designing policy in the future, the framework proposed by Horwitz and Carter in Chapter 4 may provide a useful starting point. In keeping with other chapters in this book, Chapter 9 also stressed the importance of governance.

In Chapter 10, Sue O'Keefe and Glen Jones drew on the experience of the Boating Industry Association of South Australia and the National Marine and Maritime Association in the United States to provide examples that demonstrated the importance of political suasion in achieving improved outcomes for the recreational boaters. These examples highlighted the potential for the tourism and recreation sector of both organized strategic action that harnesses like interests and the more opportunistic approach to the building of alliances.

This section was drawn together in Chapter 11 by Ronlyn Duncan, and several lessons are apparent. First is the understanding that all water allocation decisions occur in a political as well as economic space. This finding echoes the conclusion of Dollery and O'Keefe in Chapter 6. Second, Duncan showed the importance of the production of knowledge in policy formulation. She proposed a model of coproduction that offers potential advantages for the tourism and recreation sector.

Duncan argued that there is a better way for this sector. She drew on the work of Cash et al. (2006) and Jasanoff (1990, 2004) to support her proposition for a coproduction knowledge governance model. Central to Duncan's argument was the notion that the way knowledge is produced affects what knowledge is produced, and that this in turn affects the willingness and capacity of policymakers to act. Arguably, this is a lesson for all water-using interests that compete for the resource.

The upshot is that if the tourism and recreation sector is to secure more influence at the policy table, they need to be involved in the production of knowledge in collaboration with other stakeholders. This is likely to confront many obstacles, including the intransigence of those who currently hold sway in the decisionmaking process, but it is nonetheless critical if the tourism and recreation sector is to exert more influence in the allocation domain.

POTABLE WATER AND TOURISM

The fourth question is, what are the implications of tourist water-using behavior on water infrastructure, such as that embodied in urban water supply? Most attention in this book concentrated on surface water and its relationship to tourism and recreation, whereas in Australian policy circles, the focus has remained on the nexus between the environment and other users of the resource. In the final part of this book, we sought to offer a different perspective by turning attention to the impact of tourism and recreation on urban water and its supply. Central to this were consideration of the water-using behavior of tourists and the attitudes that commonly are associated with vacationing. In general, tourists place excess demands on the water resources of many towns, and this also has implications for infrastructure development.

Chapters 12 and 13 examine the causes and impacts of the increased consumptive demands that the tourism and recreation sector places on water resources. In Chapter 12, Bethany Cooper noted that the water use behavior of tourists is often much more profligate than that of locals, and that even those who are eager to conserve water at home may not be so frugal while on vacation. Cooper posed a number of questions about the motivations, attitudes, and behaviors of tourists and put forward a framework for considering motivations to comply with water conservation aims.

Lin Crase and Bethany Cooper also focused in Chapter 13 on consumptive use of water and highlighted what this means for water-providing entities, noting that the nuances of tourists' water-using behavior have important pricing, infrastructure, and policy implications. They examined a range of water-rationing policies as they relate to tourists and vacationers in urban settings, arguing for more explicit and overt pricing signals.

These authors raised several concerns about the current approach to the impact of tourists on urban water supplies and suggested some ways forward to achieve improved policy outcomes. In this regard, the compliance framework presented in Chapter 12 should serve as a useful tool to assist in the design of compliance regimes. Put simply, policymakers wishing to maximize compliance with minimum costs could benefit from knowledge about the individual motivations of its market. It is likely that the motivations of tourists will systematically differ from those of locals, and a compliance regime that contains different tools for each segment of the population may well be more cost-effective.

Chapter 13 also considered the relative strengths and weaknesses of various demand management approaches. Building on the recognition that tourists are likely to respond to different motivations than locals, Crase and Cooper questioned the efficacy of current approaches to demand management. In particular, they argued that the costs and effectiveness of a system of moral suasion and rule-based approaches may not be effective in constraining tourist demand. They also criticized current pricing regimes that are constrained by regulators to consider a range of sometimes competing objectives, such as revenue surety, cost reflectiveness, and equity. A central problem is that price signals are muted for tourists, who have little moral or social motivation to comply with the need to conserve. In this context, a more nuanced approach would seem preferable, and the use of new technologies, such as smart meters, appears to offer an alternative.

CONCLUSIONS

Some headway has been made in the valuation of water for tourism and recreation, and this bodes well for continued scrutiny of the range of complementarities and conflicts inherent in the various uses of the freshwater resource, many of which were examined in previous chapters. However, a shared understanding of the trade-offs implicit in a given water management decision is unlikely to be easily achieved. This is due in large part to the politically charged nature of water policy in Australia (and elsewhere, for that matter). Accordingly, the current absence of tourism and recreation interests should not be seen as reflective of a lack of value, but rather it is likely to emanate from previous policy choices and public perceptions coupled with the fragmented nature of the sector. The ecosystem services framework proposed in Chapter 4 offers a practical way to bring the disparate interests to an understanding of the trade-offs inherent in any water management decision. Of course, this approach requires testing in practice.

More knowledge is required, and it needs to be more productively harnessed. As Ronlyn Duncan argued in Chapter 11, the production of this knowledge is itself a political process. A recurring theme throughout this book has been the scope for more collaborative, devolved decisionmaking that harnesses the coincident interests of the parties to negotiate the conflict. Collaborative models of decisionmaking, such as that evidenced in water trusts in the United States, appear to offer some promise, but additional empirical work is needed to establish their potential in an Australian setting. Moreover, the current property rights regime that values only the volumetric dimension of water would appear to stymie the scope for the sector to take advantage of important complementarities, particularly in relation to the timing of flows. Once again, this is a question best investigated empirically.

In the context of significant policy reform in Australia, the time would appear ripe to further investigate some of these thorny issues, but this is not to say that the Australian experience can simply be transferred to an international setting. Rather, from an international perspective, we urge careful consideration of the range of issues examined in this book before rushing in the direction of particular water policy reforms that might benefit the sector. As has been shown throughout these chapters, careful prior consideration of institutional and politicoeconomic factors seems likely to secure improved outcomes for the tourism and recreation sector.

REFERENCES

Cash, D.W., J.C. Borck, and A.G. Patt. 2006. Countering the loading-dock approach to linking science and decision making: Comparative analysis of El Niño/Southern Oscillation (ENSO) forecasting systems. *Science, Technology & Human Values* 31 (4): 465–494.

Jasanoff, S. 1990. *The Fifth Branch: Science Advisers as Policymakers.* Cambridge, MA: Harvard University Press.

———. 2004. Ordering knowledge, ordering society, in *States of Knowledge: The Co-production of Science and Social Order,* edited by S. Jasanoff. London: Routledge, 13–45.

Ostrom, E. 1990. *Governing the Commons: The Evolution of Institutions for Collective Action.* Cambridge, UK: Cambridge University Press.

Index

For Product Safety Concerns and Information please contact our EU
representative GPSR@taylorandfrancis.com
Taylor & Francis Verlag GmbH, Kaufingerstraße 24, 80331 München, Germany

www.ingramcontent.com/pod-product-compliance
Ingram Content Group UK Ltd.
Pitfield, Milton Keynes, MK11 3LW, UK
UKHW021614240425
457818UK00018B/562